AVID
READER
PRESS

ALSO BY JONATHAN WALDMAN

Rust: The Longest War

SAM

One Robot, a Dozen Engineers, and the Race to Revolutionize the Way We Build

JONATHAN WALDMAN

avid reader press

New York London Toronto Sydney New Delhi

AVID READER PRESS
An Imprint of Simon & Schuster, Inc.
1230 Avenue of the Americas
New York, NY 10020

First Avid Reader Press hardcover edition January 2020

Manufactured in the United States of America

1 3 5 7 9 10 8 6 4 2

Library of Congress Cataloging-in-Publication Data is available.

ISBN 978-1-5011-4059-4
ISBN 978-1-5011-4061-7 (ebook)

In memory of Estera

Be not the first by whom the new are tried,
Nor yet the last to lay the old aside.

—ALEXANDER POPE

[P——] has invented a machine that will revolutionise the art of building . . . it will lay bricks with all the skill of the most accomplished bricklayer, with perfect accuracy, and with a rapidity that discounts the human hands . . . it will do the work of a dozen bricklayers in the course of a day, carrying up a wall as if by magic . . . the machine can be easily and quickly regulated so as to skip wherever it is desired to leave doors and windows, doing this work with seeming human intelligence . . . the machine is not complicated, will not easily get out of order, and is in every way a practical and useful invention, sure to come into general use by contractors.

—NEWS REPORT OF A BRICKLAYING MACHINE, 1905

Contents

1 - SAM

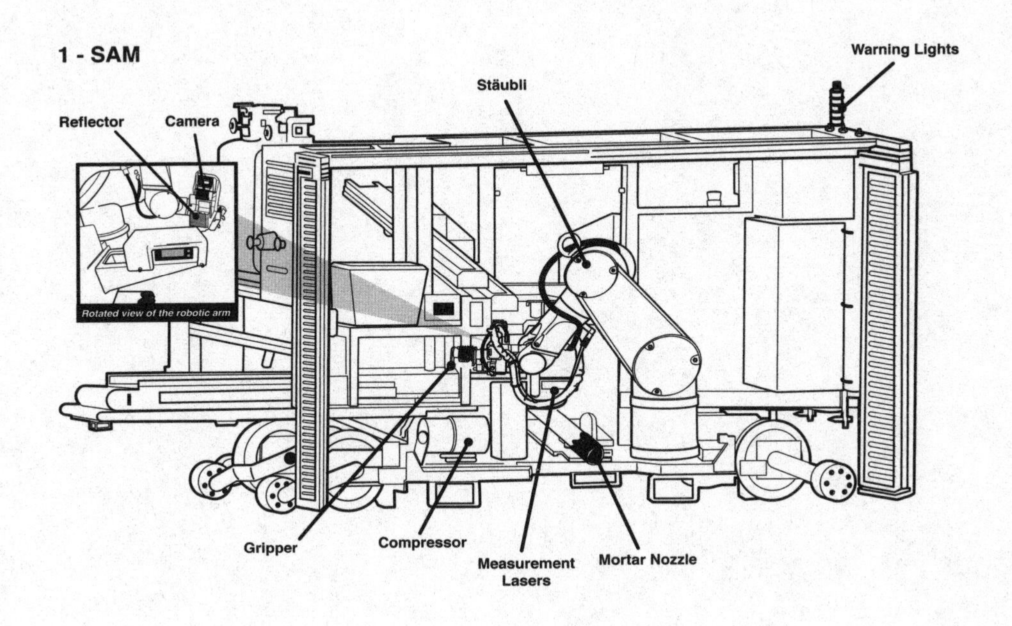

Warning Lights
Stäubli
Reflector
Camera
Rotated view of the robotic arm
Gripper
Compressor
Measurement Lasers
Mortar Nozzle

2 - SAM

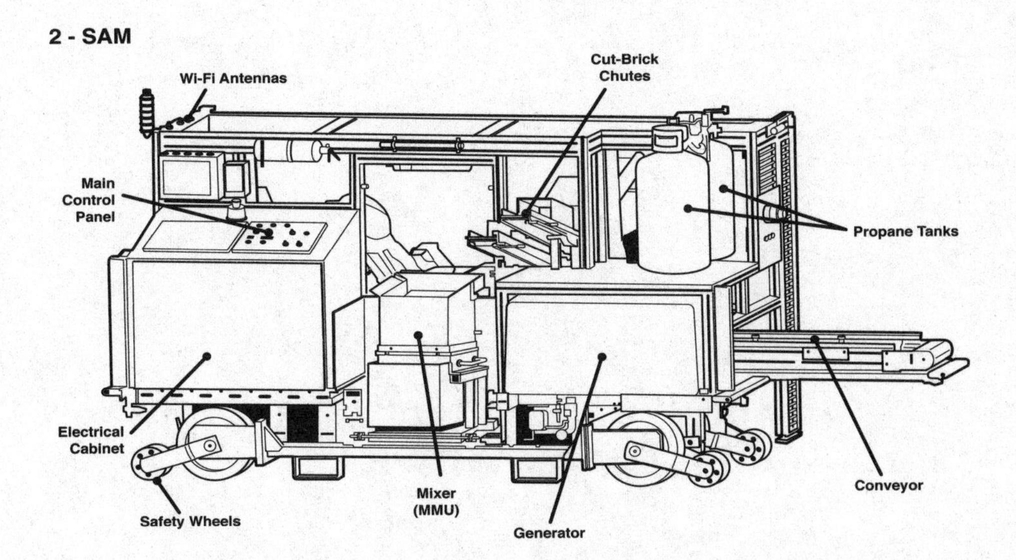

Wi-Fi Antennas
Cut-Brick Chutes
Main Control Panel
Propane Tanks
Electrical Cabinet
Safety Wheels
Mixer (MMU)
Generator
Conveyor

SAM

Prologue

Though the invasion began from both coasts, it proceeded unnoticed for years. For most of two decades—amid the Cold War and the riots and Vietnam and stagflation—it advanced in obscurity. An automobile factory was taken over here, a candy factory there. But by the 1980s, the robot invasion was unmistakable. Robots were everywhere. They were on lunch boxes and T-shirts, in comics and video games. Nintendo made one. Radio Shack sold a couple.

Across movie and TV screens, robots were veritably plastered. Mork brought one home for Mindy to meet. Mister Rogers visited some. One, named SAM, rolled onto *Sesame Street*, and insisted it was on Mulberry Street. Notwithstanding the pallid, humorless one in *Star Trek*, they had personality: Those in *Sleeper* kvetched, the squat one in *Star Wars* sassed, and David Hasselhoff's, in control of a badass Pontiac Firebird, had an ego as soft as French cheese. Connected to a vast network, the one played by Arnold Schwarzenegger was powerfully sinister; while, disconnected and on the run after being zapped by a lightning bolt, Johnny 5 was alive. So, in the end, was RoboCop.

Off the screen, they behaved more rudimentarily. One walked, one rolled down hospital hallways, one vacuumed, and another climbed stairs. In Atlanta, one "sang" at a lounge. Yet others presaged a brave new world. Two robots investigated the contaminated reactor at Three Mile Island, and another, which had found vents deep on the ocean floor, also found the *Titanic*. A NASA crew used a robot to knock ice off the side of the shuttle *Discovery*, and the army employed a robot to load hundred-pound howitzer shells. A forty-thousand-dollar Swiss gizmo helped cops in Texas dole out speeding tickets. In Maryland, where I was then growing up, a stubby fiberglass one delivered a commencement address. "Computers," it said presciently, "are now a part of us all."

I already knew as much, because my father had written a whole dictionary of robotics. It had two thousand entries (for terms like "Denavit-Hartenberg matrix" and "Karhunen-Loeve transformation") and cited no fewer than two dozen robot-programming languages, because there were more of them than native tongues of North America. At IBM, roboticists used Emily; at Stanford they programmed in Noah. At GE they used Help. Dad's dictionary explained biquinary systems, and bang-bang machines, and even a processing conundrum known as the "deadly embrace." Robotics was hot stuff. And more absurdly named than the languages were the robots themselves: the Locoman, the Pragma, the Nebot; the Grivet-5, the Prab 5800, the Cyro 2000. It was enough to make an innocent suburban boy think Voltron was real.

Four hundred miles north, in farm country east of Buffalo, New York, a boy the same age was similarly fascinated. He was obsessed with toy robots—in particular, Transformers, which did noble work and blended in with society. The little plastic toys came with tech specs, which suited a kid already enchanted by calculator watches and remote-control cars. This boy, though, was the descendant of some very inventive and courageous and stubborn men, and he had bricks in his blood.

Twenty years later, by which time the kid had become an engineer and Roombas had invaded America's living rooms, he was posed a simple question: What did he think about a robot that laid bricks? At the time, two sophisticated robots were exploring the surface of a distant brick-tinted planet. But what initially came to mind was some amalgamation of the nicely packaged, articulate, intelligent contraptions he'd absorbed as a boy. Something faster, smarter, tougher, more responsive, more precise, and more capable than humans.

He had never laid bricks. He had never built a robot.

He said, *I think that's a good idea.*

PART I

VICTOR

1.

The First Brick

By the fall of 2013, Scott Peters had spent nearly a quarter of his life anticipating a brick. It was the first real brick, and at a quarter before noon on the last Friday in October, on a west-facing wall in a suburb southeast of Rochester, it finally went down. There had been other bricks, but they were mere practice bricks, not permanently bonded components of an actual building. To Scott Peters, the first *real* machine-laid brick—not just his first but the world's first—seemed monumental, so on the deck of a steel scaffold, he squatted, removed a glove from his right hand, held out his iPhone, and captured what people there once called a Kodak moment.

Scott, thirty-five years old and over six feet tall when upright, was wearing jeans and leather boots and a mud-smudged green jacket and, to insulate himself from the cold of western New York, a warm blue beanie beneath his white hard hat. Beneath the beanie was the same close-cropped swimmer's haircut he'd had since childhood, and beneath the haircut and a scruffy beard was a boyish, toothy, open face. As the first brick went down, he was too focused—and too exhausted—to smile. Without saying a word, he pressed record.

He captured the motion of a unique contraption. Sitting on a pair of vertically aligned roller-coaster rails, it resembled neither vehicle nor construction equipment. Ten feet tall, it loomed over everyone on the scaffold like an elephant. From one of its sides, an electrical cord, an air line, and a water hose ran to the puddled ground. From another, a mechanical crutch—a peg leg, basically—pointed down. There was cardboard taped to part of it, a heating pad wrapped around part of it, and mortar slowly leaking from another part. A massive steel cabinet hid a tangle of circuitry. Stylewise, the contrivance had a lot more in common with a tree house cobbled together by ten-year-olds than an iMac or even a minivan. Two screen gates—more or less chicken wire—ensconced the thing, and a rolled-up silver tarp covered it.

Under the tarp, a gargantuan articulated silver arm, made in Switzerland, began to bend at the elbow. Above the elbow, on the arm's bicep, was a white sticker that said CONSTRUCTION ROBOTICS. Below the elbow, there was a mechanical claw that grabbed a brick from a seesawish table at the end of a conveyer jutting out from the machine's left side. The arm swung around and brought the brick to a central plastic nozzle. The machine hissed for five seconds and squirted mortar onto the bed of the brick.

The brick was a utility brick, 3⅝" x 3⅝" x 11⅝", made in Ohio by the Belden Brick Company of clay dug from deposits laid down in the last ice age. It contained five square holes and weighed ten pounds. It was unremarkable, and yet . . .

The arm twisted and extended beyond the edge of the scaffold, across two wooden planks, and into the crisp autumnal air. It descended toward the left side of a short factory wall and, as it did so, rotated the butter side down. Some mortar fell off in dribbles. As the arm lowered toward a band of white stone, it slowed as if coming in for a landing, and two red laser dots appeared beside the gripper. "Holy crap," someone said. At touchdown, there was a hum, like the first half of a siren's wail—and then the gripper opened, and the arm rose. Someone else yelled, "Woooo!" The brick remained where it had been placed, one small part of an otherwise good-looking, weather-resistant, durable edifice.

The whole movement—from pick to butter to place and back—took fifty seconds, which was, as far as paces went, far from record-setting. A human mason could have picked, buttered, lit a cigarette, taken a drag, shot the shit, scratched his ass, kept tabs on his foreman, looked out for OSHA, and still placed his brick before fifty seconds had elapsed. The world's fastest bricklayer could have placed a couple dozen bricks in that time. Scott Peters, an engineer so persistent that he'd never put the word "that's" before the word "impossible," said nothing. He had aspirations far beyond fifty seconds and dreams that involved much more than short factory walls.

He wanted to revolutionize construction, the second-biggest industry in America.

But first, he attended to the second brick, because after fifty seconds of glory, his five-thousand-pound bricklaying machine was stuck.

In the scheme of things, the inauspicious debut was a predicament of minor technical consequence, and yet a lot hung in the balance—not just for Scott but for America. Brickwork, having endured decades of decline, was disappearing. Who cared about bricks? Nobody. And everybody.

Bricks strike a sociological nerve, presenting a familiar, comforting fabric in our lives. Bricks make schools feel school-like and churches church-like and factories factory-like and banks bank-like and firehouses firehouse-like; they make institutions feel institutional. Bricks give neighborhoods, and cities, and whole regions of the United States their character, to say nothing of what they grant to European countries, where bricklaying reached its greatest heights and from which American brick masonry traces its origins. No other architectural material registers so evocatively, so naturally. Think of Williamsburg, or Fells Point, or Pioneer Square, or Chestnut Hill, or Lincoln Park, or Whittier, or the North End: Chances are, walls of baked clay units come to mind. Plop a New Yorker in Kathmandu and the Nepalese bricks cast a spell over the foreigner, suggesting he's not so far from the Big Apple. Clay resonates. Around the world, across religions, mythology has it that God fashioned mankind out of clay. "We are the clay," Isaiah instructs. Like us, bricks are of the earth; like us, bricks breathe; and like us, each brick is imperfect but also good enough.

Bricks also bear great historic significance. On some level, since the time of Noah, bricks have become unconsciously in us, of us. In a time when the phrase "brick and mortar" evokes the quaint Main Street past, it's easy to forget that brickwork has a lineage so long it's been called aristocratic. Bricks, Shakespeare knew, testified for generations.

"Kingdoms are clay," he wrote. The greatest city on earth owes its existence to clay, and bricks resurrected American cities from coast to coast when other materials proved unworthy. It was a brick wall that withstood the breath of the Big Bad Wolf, and it was bricks that this nation's earliest settlers, in Jamestown and Roanoke, set to baking immediately upon their arrival. To manage water and fire, you need bricks. As one old-time brickmaker put it, everything made by brick becomes an everlasting monument, revealing "in nature's eloquent tongue of silence . . . the modest virtues and worth of the maker." Only a fool would dare praise vinyl siding in such terms. Ever stylish, bricks command respect that the prefabricated panels of America's strip malls do not. The way the old brickmaker saw it, "but for clay, the world would be an arid, lifeless waste." A stretch, perhaps—but a brickless civilization, for all its slickness, would in fabric and texture also feel alien, and somehow betray our humanity.

Rising labor costs and declining productivity, though, were turning builders away from the world's most universally available building material (superior in strength, durability, environmental impact, and performance) and toward materials—vinyl, aluminum, glass, and steel—that could be put up more quickly (and hence cheaply). Without an overhaul in the trade's standard way of business, the teetering brick industry was at risk of fading into oblivion. It was true that good bricklayers were harder than ever to find, and that even the fastest one, when weighed against the march of progress, seemed slow, but the art of bricklaying, as ever, revealed a crucial metaphysical truth: Even our grandest aspirations require piling up innumerable small units in accordance with the law of gravity.

In this way, the business opportunity that Scott Peters saw in the decline of bricklaying was a great deal more than that—but being stuck after placing just one brick left him no closer to addressing it. Ironically enough, Scott had brought the present predicament on himself. More ironically still, Scott's engineers had warned him. His chief engineer, in fact—a lively mustachioed tinkerer named Rocky Yarid—had spent a good portion of the previous five frantic months lambasting Scott for pursuing quick and dirty engineering solutions. "People don't remember the quick," he'd said, "but they remember the dirty." Scott, though, had insisted on speed—partly due to his nature, partly due to the wisdom he'd picked up from a particular start-up book, and partly due to finances, most of which had their origins in the bank account of his gray-haired co-founder, Nate Podkaminer. With only so many funds, Scott had been pushing to innovate quickly—more quickly than Rocky or his five colleagues could handle.

The six engineers did not lack experience. They'd designed and built houses, musical instruments, windshield wipers, multimillion-dollar X-ray-film factories, and the laser recorder with which Disney digitized *Snow White*. Three had come from Kodak and, between them, had enough years on the job that they could recall not just when the company transitioned from pension plans to 401(k)s but when the buildings where they once worked were blown up live on TV. Two others, like Scott, had come from General Motors. In the factory where GM once made carburetors for Cadillacs and fuel injectors for Corvettes, GM had been designing fuel cells. Fuel cells were supposed to be the future. But twelve years and $140 million were not enough to bring about that future, and a year earlier, the fuel-cell lab had been shuttered.

These refugees of major American industry knew their young boss was smart and ambitious and uncompromising, and they appreciated Scott's disregard for meetings, documentation, and hierarchy—but they also found his approach overly aggressive. The way Scott saw it, if nearly one hundred thousand airplanes could land squarely on U.S. runways every day, how hard could it be to put a fraction as many bricks squarely where they belonged? He wanted to get building hastily. To his engineers, important things—like wisdom and finesse—were being lost in the rush. Then again, they also knew that a bricklaying machine like the behemoth before them could never have been built at a place like Kodak. Nimbleness might have caused headaches, but it begat innovation. "Quick and dirty" was what got Americans in space, after all.

The headache registered loudest to Rocky, because his boss had committed to using the bricklaying machine on the job before the machine had even been assembled. Faced with such pressure, Rocky had come up with a refrain. "Mr. Peters," he kept saying in faux formality, "you can't make a baby in three months with three women." What he meant was: Scott's approach to product development, reliant on an all-success schedule, was untenable. He was engineering too fast.

Scott's radical approach was what led to an oversize contraption—capable of laying forty-pound cinder blocks as well as four-pound bricks—powered by an undersize motor, resting on undersize rails. Those rails had bent and bound up—preventing the machine from rolling twelve inches to the right. So after loosening the contraption's wheels and engaging the hydraulic peg leg, Rocky, in a dirty brown Carhartt onesie, repeated his refrain, put his hands out, and leaned in to the machine. With half of his colleagues, he busted his butt pushing the big, unwieldy baby to the next brick.

The bricklaying machine was called SAM, for "semi-automated mason," and by Friday's end, it seemed not so much semi as barely. Merely moving the machine from where it was assembled to the job site had been such a fiasco that Scott couldn't bear to watch. Lifting the behemoth required a massive forklift, and it took the forklift a full day to transport the SAM-scaffold-rail combo less than a quarter mile. Once in place, it took three days of coaxing—pumping mud, exercising the arm, aligning the lasers, adjusting settings, remounting components—just to get that first brick. By the time the workday was over, long into overtime territory, SAM had put down all of 108 bricks, which was only a fraction of what even the laziest human bricklayer typically laid.

For Scott, the number was hard to digest. For eight years, he'd dreamed of a machine so dominant, so refined and widely dispersed, that it would render today's antiquated hand-laying technique obsolete.

In many ways, such a shift was to be more significant than the much heralded move to self-driving cars, because horseless carriages have been evolving for a hundred years, thanks to armies of engineers and billions of R&D dollars. Bricklaying hasn't changed since man crawled out of the muck. As ever, laying bricks requires hard work and a lot of time. Old-timers know that bricklayers lift the equivalent of a Ford truck every few days, basically trading a body for a paycheck. That trade frequently results in wrist damage, elbow damage, knee damage, rotator-cuff surgery, back surgery. Scott wanted his contraption to free men from that burden, and to free construction firms from reliance on bricklayers who slept in, or showed up hungover, or laid bricks slowly or sloppily or with a bad attitude. A machine that laid bricks would be unstoppable, tireless. Such a machine was . . . the future!

But placing six narrow courses of bricks on that first day had entailed such a fight that to Scott, the remaining thirty seemed like a million.

Nevertheless, Scott and his engineers climbed onto the scaffold day after day and kept at it. After two weeks, even if their rhythm wasn't smooth, they developed a feel for their machine, and had a short segment of real wall that they could point to as evidence of the machine's potential. At this point, Scott invited the world to see SAM, because his approach called for it. While SAM didn't exactly deserve to be seen, it needed to be—because Scott wanted to hear precisely what needed improving, and not just from some focus group. He wasn't a bricklayer, after all. He was a process engineer.

Also, he wanted some publicity.

For the occasion, he set up a large tent and ordered barbecue. The day was cold and drizzly but not too wet to lay brick. Dozens of people showed up at the construction site in Victor, and those excited about SAM but unexcited about the weather watched the machine from live-feed screens in the tent.

Nate, Scott's co-founder, had wanted to invite the big boys of the construction world—Bechtel, Fluor, Kiewit, Turner, Skanska, Clark, Mortenson, Yates, Suffolk—all of whom did over $2 billion of work annually. Companies like that could afford to invest two thirds of a million dollars in the latest technology and would recognize advancement when they saw it. It was probably for the best that the Big Boys were not invited and that men from ten smaller companies showed up. To Scott, these attendees still seemed like gods of the masonry world—among them a crew from Maine Masonry, a mason from Syracuse, a mason from Ontario, and eight men from Belden Brick, the country's premier brickmaker.

The visitors were not enthusiastic, but they were honest, as brick men tend to be. One, who clambered onto the scaffold, told Scott that the machine was really cool but was "not there yet." He encouraged Scott to keep going. Another, who had known Nate for a long time, said SAM was "never gonna make it." "Nate," he said in the tone of an oncologist, "you're wasting your time."

But another visitor, the owner of the building under construction, told Scott he liked what he saw, not because of SAM's proficiency but because SAM's presence was compelling bricklayers elsewhere on the job to work faster. They didn't want to get beaten by a bricklaying robot.

Uninvited, a representative of the International Union of Bricklayers and Allied Craftworkers showed up, making Scott and Nate nervous. "Who's the mason on this job?" he demanded. The BAC, which had a long history of turf battles, did not want machines taking away the jobs of union men; neither did it want jobs that ought to be theirs going to lowly operators or laborers. Delicately, Scott and Nate pointed the representative to two Syracuse masons who were working alongside SAM. A father-and-son outfit, these masons were also friends of Nate's—and believers in SAM's potential. The older one, who had inherited the business from his father, had written, "This robotic concept could be the next generation in our business. . . . You either embrace a new technology or you're left behind." In wonderment, trying to avoid getting run over or punched by the monstrous machine and also not to block its lasers, he and his son tooled the mortar between the bricks SAM laid.

This assistance was part of what made SAM only *semi*-automatic: The machine merely placed bricks, leaving beads of squished-out mortar between them. To make the wall look presentable, men still had to clean it up before the mortar hardened. Doing so entailed swiping the wall with a trowel, to knock off excess mortar, and then—with a tool that looked like a handheld lightning bolt but may as well have been a Sharpie—gently striking the joints to give them a uniform concave profile. This involved a downward motion for the head joints and a sideways motion for the bed joints. It also demanded full attention, because nearly all of the joints contained voids where no mortar had squished out. Like potholes, these had to be filled in. Had there been only a few such voids, a mason could have repaired them by pressing marbles of mortar into the wall with his thumb. Instead, he had to dab the jointer in his right hand into the pile of mud on the trowel in his left hand and apply that glob of mortar where necessary—without, of course, slathering mortar all over the faces of the bricks. The procedure was mundane and repetitive, but it was also delicate, precise, and revealing of so much human dexterity. Nate hoped someday to automate this routine, too, but did not mention the thought to the union rep.

A reporter from the Rochester *Democrat and Chronicle* showed up and, in very unlab-like conditions, interviewed Scott and Scott's business manager but not Nate—who, though his doggedness rivaled Scott's, preferred to keep his hands in his pockets and remain far from any limelight. As much from age as from experience—he'd been working in commercial construction for four decades—he was snarkier than Scott, liable to let slip an easily misinterpreted sarcastic comment, and knew it was often best to restrain himself when it came to public relations. As he put it suggestively, he'd seen a lot.

For all of the construction industry's magnitude, Nate knew it was full of dysfunction. On account of the industry's wastefulness, inefficiency, and productivity that was actually declining, the construction industry got a fraction of the venture capital investment that other industries did. The $8.5 trillion construction industry, Nate knew, was famously stubborn in its adoption of new technology, and the few advances embraced by the industry only proved the point. As such, the construction industry was not where most looked for innovation. Nate didn't say it to the reporter, but to him the construction industry was veritably begging to be disrupted.

Given the chance, a more skeptical reporter might have pointed out that you'd have to be brave, foolish, or masochistic to pursue such a path.

It was fortunate that the reporter did not talk to Rocky, because to Rocky, the whole endeavor seemed so crazy that he wondered: *What the hell am I doing this for?* For weeks, his days had started at four a.m. and ended at ten p.m., under floodlights. They ended this late because Scott always wanted more bricks laid by SAM; he wanted numbers in the triple digits and dreamed of four digits. *Bricks in the wall*, he said over and over—as if he could will it to be. And for two weeks, Rocky felt like he'd been flirting with danger. Though he had been reassured that the scaffold could handle the combined weight of the huge machine, bricks, and humans, he remained ready to jump off to the safety of the ground at any moment.

In any case, Scott did most of the talking. Though he well knew Construction Robotics' mission statement and elevator pitch, he didn't quite blurt them out. In his defense, he was new to PR, and he was overwhelmed. He wanted to say: *Just as robotics and automation are commonly found in factories today, we believe they will become standard on construction sites in the future.* In other words, the same robots that could spit out an electric shaver every other second in a Netherlands factory were steadily making their way out into the world, not just in cars that drove themselves but in fruit-picking robots, dishwashing robots, boat-cleaning robots, laundry-folding robots. In the next five years, others would emerge that mixed drinks, sliced pizza, flipped burgers, even assembled IKEA furniture. This was perhaps *the* movement of the twenty-first century—even if, among non-engineers, it prompted a general anxiety that robots were on the verge of replacing humans, of obviating manual labor and putting a vast swath of *Homo sapiens* out of work. But Scott wasn't trying to replace humans; his aim was to combine forces, and save men their jobs by marrying man and machine. By creating a bricklaying robot, he aimed to eliminate lifting and bending and repetitive-motion injuries in humans; to improve the quality of walls; to finish jobs faster and safer and cheaper; and to ease project scheduling and estimation. Basically: to modernize the world's second-oldest and most primitive trade. Surely this last quip would have done well in a newspaper.

What Scott actually said was that he aspired to bring robotics to the construction industry, and that his goal was to take what he'd learned—his "learnings"—from the contraption before him and apply it toward the next iteration. This variety of looking ahead, and of stealthily spinning failure as opportunity, was pure Scott.

Scott did a good bit more spinning—not by lying but by not revealing the whole truth.

He didn't mention the difficulty of transporting his contraption to the job site or the effort it took to get the first brick. He didn't mention that the human masons attending SAM, who were so optimistic in their embrace of new technology, had to use the heels of their trowels to tap every one of SAM's bricks into proper position, followed by a quick check with a long level, since the robot never put one down dead-nuts level. He didn't mention that SAM once punched through the wall it was building, or that his engineers, exhausted and up to their eyeballs in dirt and mortar, wanted to punch through the wall, too.

He didn't mention that it took a few hours and ten times as many curses to make a wall map (which, from an iPad, could be fed to SAM, so that it knew where to reach out and lay bricks), or that it took so long to raise the finicky guidance lasers to the next course that, during the process, the mortar in the machine's hopper inevitably hardened up and had to be scooped out before the machine went into rigor mortis. He didn't mention that because the machine tolerated mortar only so thick, one of his engineers had to constantly nurse its consistency, or that on account of the cold, the arm's transmission oil had thickened to such a degree that getting the robot to shed its arthritis necessitated predawn ignition of a space heater under the tarp, and subsequent confirmation that the whole rig had not caught fire.

He didn't mention that for the last ten months, the company's office had been an uninsulated modular trailer in a parking lot on the other side of the construction site, eight guys crammed into a drafty box with a microwave but no bathroom. Their situation was so grim that when people asked where they worked, they left it vague.

He didn't mention that because he had no idea how the world would respond to a bricklaying robot, he'd not only hired a guard to watch it but installed on it a security system that he connected to his own.

He didn't mention that his co-founder was his father-in-law, and that his business manager, Zak, was his brother-in-law (and temporary roommate)—and that, as such, his family's livelihood hinged on the eventual success of the clunky bricklaying robot.

The frustration, the extent to which he and his employees underestimated challenges, the mountainous-looking learning curve—none of this emerged from Scott's mouth. Nuances of a dozen varieties, compounded by the variable conditions inherent in masonry and construction, eluded him and his engineers—but all of this would have been difficult to convey, and Scott, ever the optimist, didn't want to spew excuses.

Stepping in for his brother-in-law, Zak Podkaminer—the youngest of Construction Robotics' employees and the only non-engineer on staff—put it concisely: "It's really not about speed right now," he told the reporter. "It's more about learning."

One week later, the pilot project came to a close less by choice than by necessity. It was snowing and the lasers couldn't beam through all the snowflakes, leaving SAM at sea without a sextant. That SAM couldn't tell where to place the next brick was actually of little consequence, because by then SAM, on account of a mysterious glitch, had grown insubordinate. It refused to place the last brick in the narrow patch of wall before it. In ten days (some days were so cold that mortar wouldn't set up, so they didn't even try), SAM had placed 1,295 bricks—for an output far less than a human's—but it would not place number 1,296. For all Scott knew, it was about to start talking like HAL and refuse to open the pod bay doors, too. The last brick went in by the hand of Rocky.

Amazingly, the story that came out in the *Democrat and Chronicle* did not convey the nature of the job. Overwhelmed by the speed of Scott's yapping (and the chaos of a construction site), the reporter wrote in his second sentence that SAM could lay as many as three thousand bricks a day. He did not use the word "allegedly," or attribute the unverified claim to Scott, or otherwise clarify the nature of this statement. Three thousand bricks a day would have taken the alignment of the stars and the favor of the gods. SAM's best daily performance was 150 bricks; the machine had come no closer to placing 3,000 in one day (let alone in three weeks) than it had in escaping the earth's gravity. But the reporter portrayed it as a fact, right there on page 7A.

In that way, Scott got some of the publicity he'd wanted. Yet the publicity did little to raise his spirits, because he was sure that the words written about SAM were not the only ones on the subject. Scott suspected that those industry leaders he'd invited, having seen his bricklaying contraption flop firsthand, were now out gabbing to other industry leaders about the machine's performance. And even Scott recognized how ridiculous that performance had been. The machine, painfully inelegant, had run so slowly that it seemed like a bad joke. The whole job seemed like a bad joke. Used to polished processes, Scott called the job disastrous: cold, exhausting, expensive, frustrating. He'd expected to feel awe and excitement, possibility mixed with ambition, but what registered as he looked up from the muddy ground was only a sense that everything had come crashing down. Upon the completion of that first wall, what Scott felt was not pride in its realization but terror that it teetered, and that the near-decade of work on which it rested was dangerously unstable.

2.

The AM Project

It was in 1995 and 1996 that the notion of a bricklaying robot had first taken root in Nate's mind, and it had happened while Nate was in Fayetteville, New York—a place already steeped in masonic history. Fayetteville was where, in 1819, the United States' first natural, or hydraulic, cement plant had opened. The cement works was created expressly for the construction of the Erie Canal, a project so grand that Thomas Jefferson—who was not exactly a small thinker—had called it "a little short of madness" and dismissed the idea as fanciful. Building the 363-mile canal and all of its various locks and gates and weirs and aqueducts had required a tremendous amount of stone blocks, but even more important, it had required mortar to hold all of those blocks together—and this mortar had to set fast, even underwater (hence the term "hydraulic"). Achieving this took cement, which—per the recipe first used by the ancient Egyptians—required a certain variety of limestone, burned and ground into a powder. Masons without limestone could burn oyster shells, as they did in Jamestown, or better yet, add volcanic ash, like the Romans. But in northern New York, masons could not get by with shells or volcanoes. In the words of one architectural historian, the quest to find suitable mortar for the Erie Canal had presented a "troublesome" problem.

In 1818, engineers in charge of the canal had tried setting stone blocks with common lime mortar—made from pure limestone—only to see the mortar quickly fail. Concerned to the point of alarm, the chief engineer suggested importing cement from Europe. However, his assistant, a twenty-seven-year-old store clerk who'd been wounded while defending Fort Erie, urged patience. He'd spent much of the previous year in England, where he walked two thousand miles of canals, all the while examining every particular of construction and material. Masons in England had been using hydraulic cement for sixty years, and the young man, named Canvass White, took note. That fall, White quickly recognized that mortar made from a yellowish gray limestone dug up near Fayetteville was different, and special.* White and his boss summoned a scientific doctor to examine the stuff's resultant mortar, and this scientist fashioned some into a ball, then put the ball in a bucket of water overnight. The next morning, he removed the small sphere, rolled it across the floor like a candlepin bowling ball, and called it a cement as good as any in the world. White and his boss soon declared it "a discovery of the greatest importance." They went on: "It sets much quicker, and becomes stronger in the air, than common lime mortar; and under water, where a common mortar will not set at all, it begins to set immediately, and in a few weeks acquires great hardness and tenacity." The rock from which it was produced was not pure limestone, but limestone whose composition was about one third clay. To the commissioners' delight, the limestone under Fayetteville (just east of Syracuse) was conveniently located at nearly the midpoint of the canal-to-be, and there was so much of it that they deemed the supply "inexhaustable."

In 1821, Canvass White patented this "water-proof cement," and within five years the state of New York had bought the rights to the patent for ten thousand dollars, so that anyone could manufacture the stuff. This was a move of great philosophical and practical nobility, because by the time the canal was completed, no fewer than twenty-eight manufacturers had produced half a million bushels of the stuff—and so many more canals were being built that Canvass White would never again lack for work. In fact, the cement that had been used to build the Erie Canal, which made the city and state of New York an economic force (and opened up the Midwest), was subsequently used to build

* It's only fitting that the name of the man who mixed up the first batch was Mason Harris.

one hundred and fifty other canals. Natural cement was used for the base (behind the granite) of the Statue of Liberty—the largest concrete thing then built in the United States. It was used on the U.S. Capitol, twelve state capitols, fifty-one forts (many of them brick), and the (largely brick) water systems of New York, Philadelphia, Boston, and Washington, D.C. It was used in the Brooklyn Bridge, the Smithsonian Castle, the U.S. Treasury, and Alcatraz. Most noticeably, hydraulic cement was used in the bottom third of the Washington Monument, which, all these years later, may as well be thought of as an archaeological artifact slowly sinking into a swamp, its destiny that of a fossil. But it's a more dispersed archaeology that remains even more impressive: In the century following Canvass White's find, half the country's cement was produced from the limestone buried in Fayetteville. Say what you want about Syracuse, but woe upon the man who impugns the rock beneath it.

Nearly two centuries after Canvass White, Nate Podkaminer had been overseeing one of the larger construction projects of his career: a five-story 166,000-square-foot medical center. It was a substantial building of terraced polygons, in its broadest form reminiscent of an early Roman temple, thanks to a central pediment supported by a pair of pilasters. To lessen the force of ancient architectural themes, the features had been thoroughly modernized: The pediment was more of a cantilevered curve freed of both frieze and architrave; the pilasters bore neither fillets nor flutes, let alone volutes or capitals; and the portico below was so shallow it wasn't really there—and in any case, it was sheathed in a panel of glass three stories tall. To either side, the brickwork was humble enough not to draw attention but not so humble as to be plain—this chameleon-like adjustability being probably the most glorious of all the traits attached to brickwork.

Throughout the medical center, all of the bricks were laid in the simplest and most common bond pattern: the running bond, in which each brick sits symmetrically atop the two below it. Moreover, the flat walls went up free of traditional ornament. They bore no indented or jutting-out corbels, no elaborate dentiled cornices, no showy lintels, no interlaced quoins at the corners. Instead, the walls were built of two shades of brick, each color bonded with mortar of a different tint. On each floor were nine courses of red bricks laid up with dark mortar, then—at window height—fifteen courses of darker maroon bricks laid up with white mortar, and then another twenty courses of red bricks and dark mortar. The darker bricks lent uniformity to the nearly unbroken row of windows on every floor, and they gave the building a striped continuity. One flourish provided even more: On the corners of every floor, in the centers of the maroon brick bands and inset a few feet from the actual arêtes, masons placed faux terra-cotta plaques, each a square as wide as a brick, bearing a four-pointed star in bas-relief. More than anything, these plaques hinted at the ebony pegs in the woodwork of Greene and Greene and made the building look even sturdier, as if its walls were joined by dovetails. Putting up all of this masonry was a tremendous amount of work, and over many months, the bricklayers of Hopkins & Reilly Mason Contractors, Inc., working for Hueber-Breuer Construction, did it on land once peppered with lime kilns, not half a mile from the old stone aqueduct that conveyed the Erie Canal over Limestone Creek.

Nate, months shy of fifty but still carrying the trim build and manner of the collegiate wrestler he'd been, was a project manager with Hueber-Breuer, a long-standing construction firm that built hospitals, dorms, offices, factories—institutions meant to last. He'd joined the company just after getting his degree in architecture at Syracuse*; in fact, he had been the first non–family member to join the firm, such was the impression he made. Since then, Nate had worked around Syracuse on buildings as high as ten stories but generally stayed busy on structures two or three stories tall. No matter the height of a project, though, the level of urgency remained constant. Delays, whether the result of logistical complications, human screw-ups, or acts of nature, were the norm in construction, and Nate's job was to keep things moving and satisfy a building's owner that he was getting what he'd paid for (or wouldn't have to pay more for the thing he'd already bought).

This was never a simple task, but if anyone had the makeup for it, it was Nate. Having grown up sixty miles north of Manhattan in rural

*Nate had studied chemistry as an undergraduate but hated lab existence so much that he turned to architecture. "Whaddya know about architecture?" his father had asked in 1969. "Nothing," Nate had said. He just didn't like chemistry.

Brewster, he was mechanically minded and resourceful and energetic: the kind of guy who made platoon leaders very happy. The son of a car mechanic, he'd always enjoyed building things and was not what his wife later called a "sitting-around person." A math whiz, he made decisions quickly. His memory was nearly photographic.* He had an innate intelligence which might have been fashioned after Casey Stengel, the brilliant Yankees manager. A perpetual admirer of speed and action, he grew restive just being around people who spun their wheels. He loved speed. When Nate was getting his architecture degree, he refused to idolize any particular architect, reserving all idolatry for Mickey Mantle, who, during his youth, had raced to first base faster than anybody. Speed was key—what else was there?†

By nature Nate could not just look but *see*, and by disposition he was a concentrator rather than a multitasker—naturally drawn, for example, to fine woodworking because of the intense focus required. And by the 1990s he had the example of his older brother, Bob, who had moved to California in the 1960s and prospered thanks to a string of entrepreneurial innovations. At a time when paper prevailed, Bob digitized track and field results, and after that he digitized the television-audience assessments compiled by Nielsen. So when Nate saw men on construction sites not fully focused, not working smartly, not employing their tools or materials or positions efficiently, he grew irritated.

Remedying such irritation was of far more interest to him than obtaining an apology. "You don't need to apologize," he once told his son Zak, who had somehow disappointed him, "just don't do it again." In another context, Nate once explained, "I just can't handle stupidity." By his own admission, he was impatient, and it was this trait that often manifested in wise-guy comments of a type that emanates only from a certain breed of New Yorker. Asked how he spelled his surname (of Russian origin), he once said, "The way the IRS spells it." Asked where he grew up, he responded, "I don't know if I ever have." Asked where he

*Nate's memory was so powerful that it sometimes frightened him. During scary scenes in movies, he shut his eyes, so as not to permanently "record" the terror.

†Unfortunately, Mantle had been a wreck, so that speed was rarely available—and if Nate had been the kind of guy to consult a Ouija board, he might have seen an omen in this.

was, he once said he was "at the corner of walk and don't walk." Live long enough, he said, and that's where you end up.

Multitaskers especially irritated Nate. "When I see people multitasking," he once told me, "it raises yellow or red flags. I see 'em as a fifty or sixty percent person." Ever the wrestler, Nate could quickly turn bossy, get in someone's face, and declare the need for some task to be performed at a higher percentage. The words came from a man whose physical presence was unintimidating: He was of average height, so clean-shaven he lacked even sideburns, sporting short, parted hair and wire-rimmed glasses. But his voice was deep, his feet were planted firmly, more often than not his arms were crossed, and as he locked his dark eyes on a target, he tilted his head forward just enough to reveal a set of eyebrows akin to those of Philip Roth.

In the case of the medical center, the pressure to *get it done* had already risen to heart-attack levels on account of weather delays. Syracuse, at least according to the *Farmers' Almanac*, boasts the worst winter weather of any city in the U.S., and the winter of 1996 was not just more severe than usual but seemed to start early and stretch on interminably. Construction ground to a halt at the end of August, as one day dawned clear and warm and the next day and nearly every one after brought rain and snow and cold. Rain fell over half of September and October, and the snow that began in November didn't relent on more than a handful of days until mid-May. In January, during a blizzard that shut down the federal government, temperatures plummeted to 23 degrees below 0. By then, so that they could get on with pouring concrete, Hueber-Breuer had been compelled to erect waterproof and heat-retaining barriers. Getting on with the construction of the huge medical center became brutal.

Through that winter and subsequent seasons, Nate stopped by the jobsite regularly, always climbing up and down the many scaffolds, looking for ways his company and all of the various subcontractors could build better. As always, "better," for all intents and purposes, meant faster. In Nate's words, he spent a lot of time on the jobsite pushing, struggling to "work aggressively." That was when Nate first wondered if there was some way to modernize the whole masonry endeavor. The notion was in the vein of Thomas Edison, who had sought to simplify construction eighty years earlier. But in Nate, the idea was so inchoate—more of a yearning, really—that he didn't mention it to his wife or kids or colleagues or anyone. Instead, it lay dormant for a decade, like, he later said, "a bear sleeping for a very long winter."

Early in 2006, stymied by another Syracuse winter and, on account of his age, not climbing up scaffolds as often, Nate drove an hour and a half west to visit his daughter Torrey, who was living and working as a kindergarten teacher in Rochester. She was his second child, a tall, sensitive, effortlessly popular twenty-six-year-old who, like her father, made decisions quickly and could grow antsy if they were not acted upon. Her boyfriend was an engineer who was just as organized as she was but, unlike her, could spend a month mulling a million possibilities in search of the optimal answer. He'd grown up, barely 140 miles west of her, in a family of uncanny similarity. Like Torrey, Scott Peters was the second of four kids (two boys and two girls), also lived near water, and also spent a lot of time at swimming pools. Where she had grown up in Jamesville, a Syracusian suburb on a gentle ridge overlooking the Erie Canal, he had grown up in Akron, a small town in farm country just east of Buffalo, below a section of the Onondaga escarpment known as Counterfeiter's Ledge, where, on a clear day, you could see the mist from Niagara Falls. They had met in the pool at the University of Rochester; Torrey had been a sprinter, and Scott had been a distance swimmer. Both regularly got in ten thousand yards a day. Before their days of collegiate swimming, they'd both swum for local teams, and both gotten shuttled around in Chevy Suburbans and cheered along by supportive parents. Nate, for his part, preferred fishing to swimming, life in a canoe to life floating beside one, but he'd dutifully gone to the swim meets.

During that visit, in January 2006, Scott, Torrey, and Nate went to get ice cream—Nate could eat ice cream any time, any day—at the Eastview Mall in Victor. There, Nate lobbed his inquiry at Scott. Somehow, Nate recognized that the logic of automated bricklaying would register with an engineer who, just like he did, focused on processes. Logic woke up the bear, because logic had a grip on Nate. Nate, who once described his role at the construction firm where he worked as trying "to make sure people don't step in the same dog shit that I did,"* had for years watched men fasten one board at a time, pour one yard at a time, place one brick at a time. In the decade since the notion first crossed his mind at the medical center job, his frustration had not relented. He thought: *There's gotta be a better way.* Maybe some kind of robot could lay bricks. He had no idea if it was possible. But medical technology seemed so advanced, while construction technology seemed so . . . laggard. So he threw the idea at the promising young engineer.

For months, they bounced the idea back and forth over email. Nate knew he'd need to talk to masons to get a full understanding of the industry. Scott knew they'd need a prototype. Nate had enough experience to know that the union would be a challenge—but Scott didn't let that bother him. He didn't read the history of the union, or scour old journals, or even check out the BAC's website; he was too busy studying the process he'd have to master. His stance was far more practical than political: When the time came, he figured, he'd talk to the union. If the union was receptive, he'd work with them. If not, he'd go to nonunion firms first. It was that simple.

Scott had begun working at GM only a few months before, after deciding that he was bored at Intel. In suburban Boston, he'd spent four years polishing chips, while elsewhere, people invented BlackBerrys. Intel's culture had been overwhelming; plus, he'd been on the night shift. He wanted a career he was passionate about and realized that alternative energy grabbed him. So he left Boston's concrete behind and returned to western New York. He applied to GM, heard nothing, then started looking into biodiesel: Soon he was convinced he wanted to start manufacturing it in his parents' garage. He sat down with Wilson Greatbatch, developer of the pacemaker, holder of 150 patents, and member of the National Inventors Hall of Fame. Greatbatch thought the future of humanity rested on nuclear fusion powered by moon-sourced deuterium—big thinking the likes of which Scott had not heard before. The Buffalo native made Scott want to be an

*Nate had long before figured out that when visiting construction sites, it was wisest to park far from the action, to avoid a nail in the tire of his car.

inventor. Then GM called, and Scott set to work on fuel cells. Immediately, his field of focus narrowed. With a couple hundred thousand GM dollars, he designed and built a machine that, using infrared cameras, determined how many droplets of water had been wicked into bipolar plates.

But a seed had been planted: In his five months off, Scott had imagined what it would be like to run his own business.

So at GM, Scott asked a colleague for advice. If you were going to pursue this crazy robotic bricklaying thing, he said, how would you do it? The colleague's answer was immediate. "Go to PMD," he told Scott. "They're the best around."

PMD, or Progressive Machine & Design, had been opened in the middle of the robot invasion by two blue-eyed nerds, Tim Lochner and Tom Coller. All but brothers, they'd met a decade earlier at Hansford Manufacturing. Tom had started there fresh from Clemson, as a mechanical engineer. Tim, a couple years older, hadn't gone to college and so got a head start as an apprentice. He became a machine builder. Tom became a project manager. After a new owner took over at Hansford, Tim and Tom ran off and started their own show. They began by building a machine for Corning that stacked glass plates at one per second. As their company grew and took on more clients, they began working with a wide variety of robotic arms. They specialized in making machines that assembled components at subhuman precision. They made machines for pharmaceutical companies, and machines for GM, and machines for Delphi that assembled and tested fuel injectors. They had an accuracy of a micron and used two dozen articulated robot arms to shuffle the injectors along. Building the first of these, as anticipated, was tricky: It took a full year and $8 million. But the copies that followed were gravy. Sixty percent of the machines that PMD made were one-offs—and none went for under six figures.* Each, as Tom put it, was an adventure.

*Years later, PMD would make the robotic machine that placed, within a thousandth of an inch, a dozen and a half ninety-pound gold mirrors on the James Webb space telescope, the successor to Hubble.

Tim and Tom, who had a hundred employees and more vertical mills, turning mills, and boring mills than you could shake a billet at, told Scott they'd be happy to design a machine that laid bricks. Of course, the Automatic Mason, as they termed it, would be a new kind of machine. Every one of PMD's gizmos was designed to work inside, bolted to the floor of a clear, climate-controlled space. The Automatic Mason would be their first machine to venture out into the wild, where nothing was fixed or level or clean—but they were sure they'd figure it out. Tim and Tom did not lack talent or experience, but their foray into this brave new world suggested nothing so much as a line from the Proverbs: "Only a fool rages in his confidence."

Tom drew the first schematic. He envisioned a robot on a gantry—a beam—that stretched between a pair of huge towers. Able to slide to five points across the beam, the robot arm could reach forty feet of wall. Rack and pinion screws would raise the beam, and a railing would protect the robot's operator (who would also have on a harness) from falling to the ground. The only manual labor on the human side would be pressing buttons. A generator, a water tank, and an air compressor would live in a truck on the ground, not far from the control system. Mortar, mixed on the ground, would be laid down via a hose in "swatches," after which bricks, conveyed up the tower and then over to the robot, would get placed. Once a course of bricks was down, the sides, or "gaps," would be filled in.

That was the vision, at least.

In hindsight, the language that Scott, Nate, and PMD's engineers used revealed their distance from the masonry trade. They called courses "rows," and cubes of bricks "magazines," and tooling "dressing." They called head joints "gaps," and thought they could all just be "filled in." They figured they'd use the "standard brick size," as if there were such a thing. Bricks come in a dozen and a half standard sizes and a thousand other nonstandard ones. Nevertheless, the engineers identified the key issues before them: picking and placing, measurement and alignment, and mortar application. Also: tooling. They figured their robot would tool the joints.

By the fall of 2006, PMD had taken to calling the design-in-progress the Automatic Mason System, or the "AM System" for short.

Mortar, engineers now figured, would not be applied to the wall en masse; it would be squirted into a brick-size form. An operator, PMD said, would align the first brick and then be pretty much off and running. PMD said they were 95 percent confident in this system, and showed they were up to the job by getting a robot in their factory to grab and place three hundred dry bricks up against a piece of plywood. Nate—who'd never seen a robotic arm before—was wowed, but Scott was not. He doubted the mortar-application approach would work.

That winter, a PMD engineer designed the mortar-form shell, calling it the Luigi Project. It was a lot like an ice-cube tray but made of 16-gauge stainless steel, coated in a layer of Teflon, and pierced by three protruding tubes so that air could escape as the form was pumped full of mud. Nobody knew if applying mortar via such a form would work, but that was the least of their technical problems. Bigger concerns remained unanswered: Would DOT highway regulations allow for the easy transport of such a large gantry? What type of robot would be up for the job? And, in what PMD labeled a potential "show-stopper," what kind of metrology system would give the robot its true position? GPS was far too slow and imprecise. Could lasers or ultrasonics be employed? It was a mystery.

At the end of the day, PMD was, as Tom Coller once put it, a company of integrators. Strictly speaking, their engineers didn't invent new technology. They took known technologies and combined them in novel ways. Ultimately, the technology to quickly tell a moving object where it was did not exist (at least not affordably), and so PMD punted, pointing Scott and Nate to the Rensselaer Polytechnic Institute. Maybe they could help.

Already, Scott was annoyed. PMD had done the proof-of-process thing for eighty bucks an hour and had said it would take a hundred thousand dollars and half a year to design the machine in greater detail. How many phases of design would it take to get a result? He wanted things moving faster—especially when he was gobbling up information at a furious rate.

Though he hadn't taken a masonry course, or even picked up a trowel to try laying on his own, Scott had read up on the principles of masonry, learning about wythes and bonds and brick sizes and placement orientations. He read *Masonry* magazine, and articles produced by the Mason Contractors Association of America, and the Portland Cement Association's "trowel tips," and technical notes from the Brick Industry Association. Having studied the properties of mortar (initial rate of absorption, bond strength) in part by reading forty pages of specifications courtesy of the ASTM, he investigated possible additives and sought companies that made mixers and pumps and even spray guns. He investigated varieties of mixers—batch versus continuous—and varieties of pumps made for grouting water wells and delivering mining-industry chemicals. He read "Mixer Mania" and combed through old patents. And because he wanted to solve the thorniest riddle holding up the robot on a gantry, he read articles on laser scanning.

But when Scott and Nate asked another local firm to build a prototype of PMD's preliminary design, the firm said it would cost eight hundred thousand dollars. It was around this time that the gantry scheme faded away.

New hope lay in the idea of putting a robot on the end of a mobile excavator. The concept was Nate's, and it stemmed from a fierce determination that lay far below his impatience, in his bones. Nate was a man of resolve.* In any case, Tom soon drew up a plan. The machine they needed already existed: A mobile excavator, with its telescoping boom and large rugged wheels, was made for the rough terrain of construction sites and cost about $150,000. Everything about the idea seemed winning.

Scott and Nate examined the capabilities of excavators made by CAT, Volvo, Komatsu, Terex, and Gradall. The Gradall XL 5100, able to hold five thousand pounds and reach thirty-five feet, seemed best. The more Nate looked at it, the more he liked it. In March 2007, Nate said he was 95 percent confident that *this* was the right approach to building a bricklaying robot capable of working on construction sites. He was also 100 percent confident that someday the robot would tool the mortar.

*He was also a creature of habit. He bought four Buick LeSabres in a row—long after Buick stopped making the model.

While Scott liked the overall robot-on-an-excavator idea, he remained wary of the particulars of mortar delivery and application. The mortar mold seemed iffy. Nobody was sure whether a brick would be brought to the mold or the mold would be brought to the wall—leaving the question *Two robot arms or one?*—or if it was even possible to squirt mortar on a brick and not have it all fall off. Scott was not a mason, but he knew this much: Masons do not build walls by applying mortar to bricks as one applies toothpaste to a toothbrush. Generally, from a bucket or a board, they take several trowel scoops of mortar and lay down a long bed of mud—enough for at least half a dozen bricks—and then place bricks on that bed one at a time. As the first brick is pressed down to the proper height (as determined by a string line), some mortar smooshes out, and masons scrape it up and deftly swipe it onto the heads of the next brick. Scott wondered if maybe this radical full-scale mortar application could be imitated by a machine—perhaps at an angle. "The key," he wrote, was the development of a "consistent and reliable applicator." He said the only way to figure it out was to build a prototype.

Presciently, Scott started thinking that the creation of a bricklaying robot would entail mastery of three areas: mortar, the machine, and software. He knew he'd need some way to instruct the machine where to go and what to build, so he jumped ahead and started thinking about buttons, commands, screens. He had his doubts that PMD could write this kind of software.

In March, engineers at RPI's Center for Automated Technology and Systems began examining the new idea, but they had as many doubts about a mobile excavator as they did a gantry. A robot on an excavator would be limited by the reach of the excavator's boom, so it wouldn't be able to lay bricks past a third story, and in all likelihood, anything extended that far would wobble like a diving board. Moreover, maintaining stability at such a height would require putting down outriggers, or wide feet. Moving along a wall would require raising and then replanting the outriggers—and this would drastically slow down bricklaying.

Now everyone was worried about speed. As if Nate's impatience had rubbed off on him, Scott wanted speed in the future as well as the present. From RPI, he wanted a metrology system schemed up by April. From PMD, he wanted mortar and a controls system figured out by May. (He wondered: Would a computer control just the robot arm or the excavator's boom as well? Would bricks arrive in cartridges?) He wanted PMD to have his first prototype built by July.

As they worked, so did he. By mid-May, Scott was convinced that the key to mortar lay in dispensing beads of the stuff directly onto bricks, so he had two nozzles made. He wanted mortar application to take no longer than a few seconds, so the bricklaying machine could place bricks quickly. He'd been jotting down notes about speed for a while:

> March 8, 2006: 1 brick/10 seconds—> 6 bricks/min—> 360 bricks/hr
> March 23, 2006: 10sec/brick—6/min
> May 15, 2007: 3600 10hr day

That July, for maybe the first time in his life, Scott scaled back his ambitions. In his notes from the thirty-first, he wrote: "15 sec per brick?" Note the question mark.

All the while, as Scott and Nate progressed, they maintained a level of secrecy. As the invention took form, it accrued value—and the men didn't want to give away the idea for machines that, in theory, could lay half a billion square feet of brick walls a year. So as Scott looked into flavors of robots, he told manufacturers they were intended to paint the underside of a bridge from a boat. PMD's engineers said they were working on an automated pallet loading system. When Nate talked to builders, he'd merely obfuscate, saying something like "I'm working on a new process that aids masonry contractors." And when Nate applied for a grant from the New York State Energy Research and Development Authority, he asked if the agency might keep the project secret. The agency obliged and, on the argument that brick masonry represents an energy- and environmentally friendly building style, produced seventy-five thousand dollars in funding.

By the spring of 2008, still sans prototype, Scott was back to dreaming of a ten-second cycle time, and broke it down thus:

pick brick—½ second
move to nozzle—1 second
apply mortar—3 seconds
move to place—2.5 seconds
place—2 seconds
back to pick—1 second

Most of a decade would pass before a machine of his design would operate at such a pace.

To make progress on the mortar front, Scott decided to hire an intern. When interviewing candidates, his questions were more like prompts: *Tell me about a difficult project of yours; tell me about a loosely defined project you once attacked; tell me about a project that required extra effort to solve . . .* He hired a kid who'd built a prototype of a baseball glove that reduced the impact to the hand. Since Scott didn't have an office of his own, he stationed the kid at PMD. So all summer, Scott worked mornings at GM, drove over to PMD during his lunch break to talk about mortar, then drove back to GM.

Mortar trials involved experiments with a range of nozzles, to see if the squirting method could get mortar to adhere to bricks. The process entailed pre-wetting bricks, since mortar loses water as it's pumped. Stymied by these mortar trials, Scott began looking further into various types of pumps, since consistent pressure was a requisite. Unlike glue, mortar must be applied with force. Masons don't just place mud, they *sling* it. Varieties of pumps abounded: There were rotary pumps, centrifugal pumps, impeller pumps, peristaltic pumps, progressive cavity pumps; pumps run by a drum, a piston, a diaphragm, a nutating disc. Which would work well with something as dense, rough, and sludgy as mortar? Meanwhile, Scott thought about prefilling little canisters of mortar: single-serve, if you will. With that notion in mind, he wondered if, somehow, a camera could detect laser crosshairs on a wall and direct a robot to that spot. More broadly, he wondered if a machine that sold for a hair under a million dollars would get anywhere.

By the fall of 2008, Construction Robotics had spent a couple hundred thousand dollars, and aside from some mortared bricks, didn't have much to show for it. But by then, after a lakeside ceremony almost universally attended by swimmers (including Scott's high school coach), Scott Peters and Torrey Podkaminer had gotten married. Scott was now not just Nate's business partner but also his son-in-law.

3.

Dream Machines

Bricklaying robots have entered the imaginations of many men, but few have tried to build them. Many dreamers, in a sadly parasitic manner, have taken out patents on ideas—and then sat there, like Venus flytraps, waiting for tinkerers like Scott to come along and do the work for them. Of the various contraptions cobbled together around the world, the most impressive machine Scott had seen was a Swiss creation called ROB. Built in 2008, ROB was a modern six-axis robot, on a short track, in a twenty-foot shipping container. From three conveyers, it used suction cups to grab bricks and placed them not in planar walls in the running bond but in intricate, computer-designed patterns. Its cycle time was twelve seconds, or three hundred bricks per hour. It used a modern adhesive instead of mortar, though, because, as an engineer elsewhere put it, applying mortar to bricks took "considerable dexterous motoric skill." In any case, ROB never went to a jobsite. Often, in a parking garage, it built short walls that were hoisted elsewhere once the adhesive had cured. Recognizing ROB's limitations, its devisers built a mini version on treads, called dimROB, but it looked more like a snowmobile than a piece of construction equipment.

For half a century before ROB, it was almost a requirement that budding engineers and computer scientists take a stab at making machines that manipulated blocks. Getting an inanimate machine to do what only hands and brains could was apparently some kind of universal geek fantasy. And while it sounded like child's play, it was phenomenally difficult. To put it in context: The first machine that successfully picked up small wooden blocks did so only eight years before humans landed on the moon.

The first man to compel a computer-controlled arm toward this end was Heinrich Ernst. His machine, the Mechanical Hand-1, was enormous. Its shoulder, elbow, hydraulic arm, and hand took up most of a room at MIT, as did the computer that controlled it. Smoking a pipe and drinking coffee, twenty-eight-year-old Ernst put his feet up while his machine slowly identified and manipulated blocks. It operated on rules and hunted with vision. Sometimes the MH-1 did things that made no sense, but on four out of five tries, the machine got it right, and spectators would applaud. But working with mere blocks overwhelmed him. "We soon realized," he wrote, "that a great many problems were involved, more than we could pay attention to at one time." Still, the endeavor had obvious appeal: Where computers had formerly been limited to doing math, now math translated into action. Ernst, who went on to work at IBM, saw potential for computer-controlled machines to explore radioactive zones, the deep sea, and space. He also foresaw some difficulty.

Action, of course, was what industry wanted. In Lansing, Michigan, the Planet Corporation (which would become Armax) built the Planobot to move hot castings; in Connecticut, two inventors whose past included inventing the automatic door and the microwave oven (the "Speedy Weeny") built a two-ton, two-axis arm called the Unimate. Immediately, GM bought it, and—big surprise—used it to stack up hot and heavy die-cast car parts. To produce more Unimates, Unimation was formed. On the heels of the Planobot and the Unimate, a line of pick-and-place robots called Versatran quickly arrived from American Machine and Foundry, a company founded by the inventor of the automated cigarette-manufacturing machine as well as a pretzel-making machine, which rolled and tied fifty pretzels per minute and sent them on a conveyer to be salted and baked. (Thus the great-godfather of the robot invasion: pretzels!) Those first three robots—the Planobot, the Unimate, and the Versatran—cost on the order of twenty-five thousand dollars. Though they relied on magnetic tape and patchboards, they were strong enough to lift a person, and could reliably do so minute after minute, hour after hour, for four and a half years straight. Many were employed to stack (but not lay) bricks. When, five years later,

Johnny Carson invited a Unimate on his show, he had the machine pour a glass of beer and conduct the *Tonight Show* band.

By then Marvin Minsky (who'd inspired Ernst) had flexed his might. Since high school, he'd been thinking about how minds work. At MIT, on the advice of a zoology professor, he'd started dissecting crayfish claws to figure out which nerves actuated the muscles in the claw. He learned to operate the individual nerves to such a degree that he could get a claw to reach down, grab a pencil, and wave it around. This got him interested in robots. He teamed up with Seymour Papert, a South African who held two PhDs in math and liked working with LEGO blocks. Soon, using fourteen hydraulic cylinders, Minsky made a four-axis arm, to which he connected a television camera, and taught it to catch balls. The camera, however, was more enticed by shadows and reflections, and Minsky's robot tried to catch scientists, too. Then Minsky taught the machine to stack up blocks. And just like that, he reached the limit of what was possible. As much as he dreamed of a mobile robot, he recognized that just coordinating hands and eyes presented enough of an engineering problem. Everything else he called extra and irrelevant.

A lot of the pioneers in robot development (and AI) worked with blocks, and most came from MIT. While Minsky built a twelve-axis beast called Tentacle Arm (it moved like a metal octopus), an alum named Victor Scheinman created the Stanford Arm. He soon changed the name to PUMA, for Programmable Universal Manipulation Arm, and started a company called Vicarm—for Victor's Arm. It was the first six-axis arm in the world, and it was all electric, not hydraulic. Scheinman sold the design to Unimation, which teamed up with GM until Westinghouse, and then a Swiss company that soon devoted itself entirely to robots, swooped in.*

By 1983, there were six thousand robot arms working in the U.S., generally preoccupied with spray painting and spot welding in the automotive industry. Some were astonishingly precise—threading a needle, for example—but Scheinman said the future lay in smarts and the "capability to interface with the real world." At IBM, engineers interpreting "interface" developed a robot that could pick up an egg.

*Minsky, Papert, and Scheinman all died in 2016, but GM's prototype is at the Smithsonian.

In Australia, engineers developed one that sheared sheep, and in Israel, engineers developed one that picked fruit. But Japan—which had more than 125 robot manufacturers, and three times as many robots as the United States—was the center of robotics action. While Japanese engineers did not devise a robotic bricklayer, they did get robots into rice paddies and onto construction sites. The first of those, called the SSR-1 (the Shimizu Site Robot), sprayed fireproofing on walls. It looked more or less like an arm on a golf cart—a clumsy predecessor to dimROB. The other lifted steel beams. It was called, yet again, the Mighty Hand.

So while bricklaying *robots* remained fantasies, engineers devised simpler bricklaying *machines*. At least eight men, in England, Chicago, Alabama, Georgia, and New York, had seen the potential for bricklaying machines, and devised (and maybe even patented) methods of laying down two hundred bricks per hour and more. One man claimed that his machine, the Motor Mason, laid over twelve hundred bricks per hour, and another claimed his machine ran ten times as fast. These machines made it into *Scientific American*, *Engineering News-Record*, the *Brick and Clay Record*—and not much further. "If such devices really worked," a masonry analyst wrote in 1974, "they would certainly be in broad general use today; but, none have become commercially successful."

Then Bruce Schena came along. Bruce Schena, born in 1964, was the son of a general contractor in North Andover, Massachusetts. He grew up building houses and, unaffiliated with a union, first laid bricks when he was twelve. Six years later, he ran off to MIT. He pursued engineering and robotics. In his undergraduate years, he worked in the Space Systems Lab—devising an underwater robot for NASA that could simulate building in outer space. Bruce worked on the gripper of a robot that was designed, in his words, to build anything, anywhere.

During Bruce's freshman year, MIT established the Center for Construction Research and Education in its civil engineering department. The program involved architecture, planning, management, chemical and electrical engineering, computer science, materials science, economics—nearly every department. The year Bruce graduated, the CCRE created the Program on Advanced Construction Technology, thanks to a $10 million grant from the Defense Department. It was a mutually beneficial arrangement: The DOD, particularly the Army Research Laboratory, saw commercial and military technology going in opposite directions, and wanted to be sure state-of-the-art technology kept coming its way. MIT, at the time, was home to the country's most active university patent office. Some advancements were expected to go toward the Strategic Defense Initiative (aka the "Space Fence"), but the DOD, being a gargantuan builder, also sought improvements in construction—in materials, computing, management, sensing, productivity, and automation via robotics. Bruce Schena decided to stay and earn a master's degree. In 1986, under a maverick professor named Alex Slocum, with half a dozen other grad students, he became a member of PACT's inaugural class.

America at the time was in a rough patch. For a hundred years, it had been the world's superpower . . . and then, less than two generations after the Manhattan Project, something happened. U.S. Steel imploded, and the space shuttle *Challenger* exploded. The nation's cradle of industry became known as the Rust Belt. Something was obviously wrong with American manufacturing—and it didn't help that construction costs were rising while production slowed. While Detroit spat out Buicks, Japan leapt ahead. Robotics promised a way out of the quagmire, which was why, as Bruce Schena's classmates set to work on framing robots, he set to work making a block-laying robot.

In the fall of that first graduate semester, Alex Slocum brought his students to a construction site—the State Street Bank. Bruce had seen residential construction from the inside but quickly came to realize that commercial construction was a much bigger mess. Everywhere, he saw organized chaos—mud and puddles and scraps and things falling and machines moving and people scrambling and barely a flat surface around—and now recognized that construction sites were terrible places for robots. Because they were so unstructured, construction sites were worse than the cold vacuum of space. Still, Bruce moved forward with his project. Two and a half billion square feet of concrete block walls were being built every year, and the Army Corps wanted a machine that could do some of the work.

To do so, a block-laying machine, Bruce reasoned, would live on a scissor lift. It would have to compensate for vibration. It would need software that told it where to leave window openings, and somehow it would need to dispense mortar onto a steady supply of blocks. Using a laser metrology system, it would figure out where it was and place a block every eight seconds. It would have to be repairable. The whole thing, he figured, would weigh half a ton and cost $140,000 to build. If it sold for $200,000, and worked four hours a day, it would pay for itself in four years.

Bruce Schena realized that most of these requirements were out of reach and/or impractical. He didn't build a laser-metrology system, and he knew dispensing mud would be "super hard." Mostly, he recognized that keeping his machine supplied with blocks would be as tricky as getting it to lay them in the first place. He knew he was just shifting the problem from one place to another. Building a block-laying robot was woefully more challenging than building one for outer space. But getting a robot to lay blocks would be so damn sexy! So, from designs for eleven different grippers and wrists—he was a hand man, after all—he picked the best. He connected them to a shoulder. He sized the parts based on Castigliano's energy equations. He machined all the parts himself. Forgoing mortar and the mess it would make, he had his machine lay blocks dry. Since regular blocks varied in size, he used special precision-ground blocks, made in Sweden. Lo and behold, Blockbot worked. It laid blocks side by side for no longer than five minutes at a time, but it laid them so fast it broke gearboxes. All told, Blockbot laid five or six dozen blocks. His was a mighty hand indeed.

Schena had built a partial prototype, and even though he bypassed major hurdles—navigation, dispensing, supply, the realities of a jobsite—he thoroughly laid out the requirements for a block-laying robot to achieve adulthood. Thirty years later, he said, "You had to solve the whole thing or you didn't solve the problem." He knew that by neglecting the logistics of block supply, he'd merely turned a block-laying problem into a block-supply problem—without improving efficiency. "It's enough of a trick having the mason build a wall," he said. Building a wall five or ten times faster was that much trickier. But so many people, he said, had been enamored with the sexy aspect—the machine-acting-human, which makes great videos. He called this a common affliction in robotics.

"We thought we were more ready than we were," he said. The question had been: Where did robotics fit in construction? His answer was: Nowhere. He also said, "If robotics is anything, it's incremental. It's just the way it goes." In his thesis, he put it this way: "General purpose construction robots (e.g., robots that emulate a human worker) are neither technologically nor economically feasible, and will not be for the foreseeable future."

After Blockbot, the world wanted to get in on the action, and a more practical space race began. In Germany, engineers dreamed of putting an enormous robotic arm on a tank. They called their beast BRONCO and figured it would lay a brick a minute. They ran it with a joystick, and it oozed wires. Though they dreamed of applying mortar, they wrote, "the exact application of such a layer of mortar is technologically difficult and thus not suitable for automation." (Determined to extract publishing success for engineering failure, the same engineers wrote a page called "Limits to Profitability of Automated Masonry.") The Germans came up with an enormous second machine, ROCCO, which could lift eight-hundred-pound blocks. In Israel, engineers tinkered with a machine capable of stacking gypsum blocks. The Japanese threw more R&D money at robotic wall construction. So did Korea. In Finland, engineers schemed up two bricklaying robots. One, on a scissor lift, was to use suction cups to grab bricks and dunk them in a bowl of glue. To check that the right amount of glue had stuck, it would weigh bricks before and after dunking. The Finns figured it would take $2.5 million, eight engineers, and three years to develop—and that their resultant machine would lay a brick every ten seconds. The French also could not conceive of getting a machine to lay mortar as well as a human. "The reproduction of such a skill by a machine," they wrote, "would require . . . human capacities!" You could almost see them waving robotic arms in the air. Rather than use bricks and mortar, they hoped to use self-positioning blocks—giant LEGOs.

Russian engineers from the All-Union Research Institute of Construction and Road Building Machinery, in Moscow, dreamed of a rotating machine seven times faster than the one designed by the Finns. They imagined extruding the beds and injecting the heads. But navigation and sensing stymied them. Not to be outdone, the British, at the Construction Robotics Unit at City University, London, schemed up a machine reliant on bricks specially shaped to accept mud in the heads. When push came to shove, they made do with standard English bricks, sans mortar. For positioning, they used rotating lasers but were stymied by collisions and wished it were possible to take advantage of an image-processing system. Ultimately, they figured it was wisest to render aesthetic considerations moot by plastering over any walls they built. Students at the University of Illinois at Urbana-Champaign (also funded by the army) came up with Brickbot, and students at the Construction Automation and Robotics Lab (CARL) at North Carolina State came up with ERMaS (Experimental Robotic Masonry System). All of these schemes arose in the decade after Blockbot, and almost everyone cited Bruce Schena.

This flurry of activity led many engineers to publish reports in a niche journal (the *Journal of Automation in Construction*) and, beginning in 1984, gather at the International Symposium on Automation and Robotics in Construction (ISARC). Ten years in, however, *Engineering News-Record* saw it necessary to call a spade a spade. All of these advocates of automated construction were, in its words, "persistent suitors desperate to be appreciated." Automated construction was still a fantasy.

Five years after the DOD invested in Blockbot at MIT, it teamed up with the bricklayers' union to develop a block-laying machine with more prospects. What the Corps' Construction Engineering Research Lab wanted, actually, was an extraskeletal device—but the union told them that such a contraption would make no sense up on a scaffold, where men had to duck and crawl and climb around and stay balanced. Instead, the Army Corps got MAMA: the Mechatronically Assisted Mason's Aide.

MAMA, basically, was a trolley track that clamped onto a scaffold from which an extendable arm and gripper dangled. It didn't mess with mortar—a mason still had to slop it on the wall the usual way, with a trowel—but MAMA decreased 90 percent of the lifting, turning a forty-pound block into a four-pounder. The machine was designed and built by a suburban Maryland firm called M&M Consulting. Really, M&M was two men: Miller and Mayhall. Curiously, Mayhall was a mechanical engineer at the postal service—another place where automation was making a foray.

Miller and Mayhall patented MAMA in October 1993. A year later, the union tested it in West Virginia—where proud John Henry had raced a steam drill and won, and then died. Man versus machine went better this time. The masons encountered a small learning curve but quickly discovered how to make the most of MAMA and managed to increase their productivity while decreasing their exhaustion.

Excited, the union brought MAMA to Miami, where it was holding an annual conference. The union wanted to do a little demo on the beach. Four days before the demo, Hurricane Gordon ripped through—after which MAMA's controls stopped working. Not long after, the head of the project at the union retired. That was more or less the end of MAMA.

Still, other tinkerers couldn't resist making bricklaying machines that worked as MAMA did: all meat, no gravy. An eighty-two-year-old Rochesterian, who'd been a mason in his youth, made his own machine, called the VR-2000. A college student at the University of Toronto mocked-up a similar machine called AMPRO-BR that, he said, could place twenty-four hundred bricks per hour. He might as well have said that he could flap his arms and fly like a bird. Another Canadian, who'd come up with brakes for Rollerblades and an improved umbrella, devised one, too. Perhaps seeing the difficulty in automated bricklaying, a New Mexican man began pursuing automated adobe construction.

Aside from the Swiss, though, only one other group successfully threw a six-axis arm at bricklaying. They were from Harvard, and they didn't use mortar. In a nod to Ernst and Minsky, they got a robot to place four thousand wooden blocks.

When Bruce Schena got his master's degree and left MIT, he didn't look back, because it wasn't like Blockbot begged for commercialization. The whole project, Bruce later thought, was naive. He went on to work on a robotic golf caddie (which failed) and a robot that could clean up nuclear waste (which also failed). "Robots that don't quite solve problems, that's a theme," he explained. So was attacking hard problems. Then, from 2002 until 2014, Schena was the key electromechanical engineer at Intuitive Surgical, the company that makes perhaps the world's most famous robot, the da Vinci surgical system. Doctors using the robot perform surgical procedures with less blood loss than doctors working without it. In medical lingo, it improves outcomes for patients and hospitals. You've probably seen the da Vinci robot on TV. With it, a surgeon in Miami can cut the skin off a grape in Memphis. "Finally," the incrementalist said, "the technology and the ambition lined up." Finally, in circumstances nothing like those presented at a construction site, one of his robots had solved a real problem.

4.

An Arm and a Laser

As 2009 began, Scott was still overseeing mortar tests, and they were going so poorly that he was back to wondering about the feasibility of applying mortar to the wall at all. When NYSERDA turned down Construction Robotics' application for a second grant because the start-up could not yet point to a successful demonstration, Scott wondered if he might make his brick masonry innovation seem even greener by using a battery-powered basket lift, or a biodiesel truck, or even a fuel-cell-powered truck. He continued to dream big. In his notes, he wrote, "How many brick buildings in NY?"

Engineers from RPI finally showed Scott the sensing system they'd devised, and it was so complicated that it made the simple, level string line used by masons everywhere seem worthy of the Nobel Prize. RPI's system used triangulation and computer vision and spinning lasers. The spinning lasers, fixed on one side of the building, defined two flat, parallel planes at the height of the course to be laid. On the robot's gripper, two photo sensors detected the lasers—both the time the beams hit and the angle at which they hit. In this way, it was possible to calculate position and to figure out roll, but to get yaw and pitch, the engineers said they'd need another photo sensor and an accelerometer. The lasers already had to be aligned within a millidegree and, even so, weren't especially accurate beyond twelve feet. Notably, RPI said using just one laser was impossible. All this to lay a humble brick flat!

As the snow melted, Scott dreamed of a laser tracker made by Leica, the multi-tentacled optics company popularly known as the maker of the slick, expensive camera favored by Henri Cartier-Bresson. Leica's laser tracker featured an absolute laser interferometer that, thanks to a feature called PowerLock, could automatically find its target reflector up to five hundred feet away, and quickly assess its distance within ten microns. As far as metrology systems went, it was super-sexy, way slicker than RPI's involved spinning-laser system. Scott could imagine attaching the tiny target reflector to a gripper, mounting the device itself to the top of a building's frame, and not touching a string line or a tape measure or any other laser ever again. But Leica's laser tracker cost a quarter of a million dollars, so it remained solely in his dreams.

Because he yearned for a prototype, Scott started thinking of a new design. The gantry was out, and the mobile excavator had its limitations, but what if a bricklaying robot somehow rolled along a mast-climbing platform? On a construction job near Syracuse, he'd seen such a thing and been intrigued. Not only did the technology already exist, but it was cheaper than a wheeled excavator, and there was no limit to how high it could go—some were, unbelievably, being used to build New York's Freedom Tower. And Hydro-Mobile scaffolds were ubiquitous: Eight thousand of them were scattered about the country. They were also reliably strong, capable of handling much more weight than old-fashioned frame-and-brace scaffolding—weight such as, say, a bricklaying robot.

As the mortar tests continued into summer, Scott paid nine hundred dollars for an assessment of the masonry industry and got more than his money's worth of reassurance. Ten billion dollars' worth of bricks are installed annually in the United States. An eighth of new construction in New England uses brick. A third of banks and churches and hospitals and government buildings in the U.S. are brick; so are half of schools. Of all the surface area on the exteriors of all the nonresidential buildings in the country, a quarter is covered in brick. By the end of the summer, Scott and Nate saw vast potential, and they began to proclaim their innovative undertaking more loudly. "The construction industry," they wrote, "has not advanced with technology as have other industries." Soon, they figured, their machine—their *hypothetical* machine—and five men would be able to do the work of a dozen.

Pondering all the business potential, Scott began thinking about business strategies: Sell or rent? He wondered about finding investors, and strategic partners, and about working with the union. Would the machine of his devising be seen as a tool that improved the lives of masons or as a threat to them? Inspired, he signed up for an entrepreneurship class, to begin the following year, at High Tech Rochester. All the while, Nate encouraged him.

It's worth noting here that Nate did not shy from large projects. Thirty years earlier, with the Breuers of Hueber-Breuer, Nate had bought sixty acres of a thickly wooded former tree farm southeast of Syracuse and developed a neighborhood. In typical developer fashion, the Breuers called their lane Canyonwood. Nate called theirs Kinder, the surname of his wife, Jane, a fellow architect whom he'd met at Syracuse. (A Mainer, Jane was a Red Sox fan, while Nate was a Yankees fan, but somehow they coexisted.) The families built houses on adjoining lots, which in time made it easy for their kids to play together. But first, Jane designed a large brickless one-of-a-kind house, allowing Nate to incorporate one feature of his own. Nate had long admired the 1930s glazed tile silos of upstate New York, and he wanted one in his house. He wanted his staircase in one. On a farm a few miles from Jamesville, he found such a silo, fourteen feet in diameter and thirty feet tall, and offered its owner a couple hundred dollars for it. With the help of a structural engineer, he cut it into three sections and protected each in a steel frame. Then, on a cold day in October 1978, he moved the segments to Kinder Lane. Union riggers and friends stood around watching, saying Nate had lost his marbles on account of curved masonry units. It was the hardest single day of Nate's life, and one of the most expensive, but also one of the most rewarding, and Nate, every time he walked up his stairs, never forgot that.

That fall, after talking to many robot manufacturers, Construction Robotics bought a robotic arm made by Stäubli, the Swiss company that, a generation earlier, had bought Unimation. Stäubli—pronounced *schtoy-blee*, like a cigar—was not Scott's first choice for the arm, because their robots were notoriously tricky to program. But Stäublis could handle the payload and the speed, and they had just begun making waterproof robots, dubbed HE for "humid environment." They'd been developed for the food industry—cutting sides of beef or stirring vats of cheese curds—and could be washed down. They had tough finishes and internal motors. Models were available off the shelf, which was more than the big six robot manufacturers could claim.

A company with three generations of experience making loom parts, Stäubli began distributing the world's first electric six-axis robot, the PUMA, in 1982, and manufacturing it in 1989. ("Six-axis" means six joints: a shoulder, an elbow, a wrist, and three more that humans don't have.) Three years later, Stäubli released an improved robot, the RX. The robot's controls, though, were made by Adept, which had spun off from Unimation and stayed in California. As the 1990s wore on, Stäubli and Adept realized that remarriage of body and mind wasn't working—so Stäubli developed its own controller, and Adept developed its own robots, making the companies competitors. Adept is now the largest American maker of robots, but Stäubli, whose U.S. base is on the other coast,* is bigger. In 2005, Stäubli bought the rights to Bosch's four-axis robot design and now, among all the world's robot makers, is one of the few that makes four-axis and six-axis machines. Even FANUC, the world's largest robot maker, doesn't do that. Stäubli, which does over a billion dollars of business annually, sells more than six thousand robots a year.† The machines range from under a hundred pounds to over a ton. Most are used in automotive manufacturing, but a lot are used for plastic injection molding. Because Stäubli's robots are so accurate, they're also used to bend orthodontic braces, to drill holes in bones (for, say, hip replacements), and to perform hair transplants—locking in on *one hair follicle at a time.*

This precision is possible because Stäubli, alone among robot manufacturers, makes its own gearboxes. It's no exaggeration to say

*Stäubli, based in Zurich, maintains a U.S. office outside of Spartanburg, South Carolina, along with PMD and many other European companies, thanks to a BMW manufacturing facility there.

†Adding to the million and a half robotic arms already out there.

that those gearboxes are built with the precision of a Swiss watch. By measuring every ball bearing with a micrometer, Stäubli makes joints with zero backlash. Its machines don't wobble. Customers with experience using FANUC, ABB, KUKA, Kawasaki, Yaskawa Motoman, Nachi, Denso, Mitsubishi, and Epson robots report that none is more accurate than a Stäubli. Nobody else could pull out hairs. As such, Stäubli is growing fast, at such a rate that it will soon join the ranks of the top six.

Stäubli has two thousand patents, give or take. Its products are used in high-speed trains, wind and solar farms, machining, painting, drug making, semiconductor making, label making, carpet weaving, and carbon-fiber weaving. Stäubli machines at Merck make shingles vaccines. Whatever kind of car you drive, chances are good that some part of it—the brakes, or the dashboard trim, or the airbag, or the seat belts—was touched by Stäubli. Yet everybody at Stäubli thought a brick-laying robot was crazy. It was so out there. One sales rep, who had two decades at the company, thought: *Oh yeah, that's not gonna go very far.*

The fifth year was a lonely road. Though Scott liked Stäubli's robot, he didn't like its controller and, having lost patience with the company's customer service, wanted to develop his own. (As if he didn't have enough projects.) On the argument that advancements in robotics could help in other fields, such as with the handling of IEDs, he applied for a small-business grant from the National Science Foundation—and didn't get it.

With day one of his entrepreneurship class approaching, Scott dove into start-up books. He read *The Art of the Start* and *Made to Stick: Why Some Ideas Survive and Others Die*. He read *Crush It! Why Now Is the Time to Cash In on Your Passion*, and the older *Crossing the Chasm: Marketing and Selling High-Tech Products to Mainstream Customers*. He read *The Soul of Leadership*, and *The 360-Degree Leader*, and *Delivering Happiness*, and *Call Me Ted*, about Ted Turner. He read T. Boone Pickens's book, *The First Billion Is the Hardest,* and *Going Lean: How the Best Companies Apply Lean Manufacturing Principles to Shatter Uncertainty, Drive Innovation, and Maximize Profits*. This last book—which fetishized Japanese management techniques that outshone American ones—was one of the first to advocate a certain level of chaos and to deploy the word "disrupt."

Once he was in the classroom, the word emerged much more. There, Scott wrote a sales pitch for Construction Robotics. He figured the bricklaying machine would do the work of six humans. He learned back-of-the-envelope financials and wondered what would happen when, someday, shares of his company were divvied out or what dilution would do. In other words: He got way ahead of himself, just like Ernst. He also decided he wasn't a partner but a co-founder. He talked to Nate about formally dividing ownership of Construction Robotics. Nate, in whom the idea had germinated, agreed, in exchange for holding 51 percent; Scott, at whose hands the idea would come to life, had 49.

At High Tech Rochester, Scott was introduced to a project manager at a local construction firm called LeChase, which had a long history in town. For the first time, Scott abandoned secrecy and approached LeChase's project manager to ask if the firm would be interested in a six-hundred-thousand-dollar bricklaying machine. Intellectual property was no longer a concern; Scott had realized that the bricklaying robot was less property than notion—like a rocket to Saturn—and still unpatentable. Scott tried to get LeChase to commit to using the not—yet—built machine, but the firm resisted. LeChase, however, did like the idea of a machine that worked on a Hydro-Mobile platform, because it was familiar technology. Tellingly, LeChase's project manager also said that someday his firm would be more inclined to rent the machine than buy it.

In Syracuse, Nate started doing his own market research, searching not just for potential players but for potential funders as well. To two dozen carefully chosen companies, he made cold calls. Only a quarter of them expressed interest.

The year 2011 began with more entrepreneurship studies, via a boot camp at the Entrepreneurs Network. At this boot camp Scott read papers from Harvard Business School and looked at charts of consumer-adoption rates. He learned about business arrangements like sales forces, equity, term sheets, valuation, dividends, liquidation, preference, anti-dilution provisions, redemptions rights, advisory boards, and registration. Optimistically, he conjured the size of his market at eleven hundred machines in a $10 billion industry, which amounted to a bricklaying robot on one out of every eight Hydro-Mobile scaffolds across the country. He also honed an elevator pitch. Other boot campers advised Scott to nail the language about assisting masons, and to change CR's name to something more name-like and less descriptive. Scott did one but not the other, and wrote a business plan. With it, he made it to the top five of a regional business-plan competition. Encouraged, Scott and Nate began discussing starting a real business. Where, five years before, Scott had thought the wisest course of action was to sell the patented idea to some other company, now he wanted to run with it.

Running along, Scott hired MBA students at the Rochester Institute of Technology to assess the needs of potential customers. The MBA students created a questionnaire and emailed it to architectural firms, construction firms, and masonry contractors, saving the best questions for last. The eighth question of the survey asked: "What would you say if I told you that a robot has been developed that can lay bricks with near precision at a rate <xx%> faster than a typical crew, at <xx%> of the cost?"

The ninth question: "What proof would you need to accept the veracity of that concept?"

The tenth question: "How would you determine whether or not you would buy one?"

Soon enough, the MBA students compiled their responses into a report. It did not inspire confidence.

Architects and engineers said they doubted a robot could really lay bricks, especially in the cold. Would the robot's joints meet water-intrusion and strength specifications? And if they did, wouldn't perfectly laid walls look strange? Still, architects and engineers said they'd welcome any system that lowered the cost of labor.

Contractors looked inward and admitted that they tended to resist big innovations, generally operating with "show me" attitudes. And masons—whether they carried union cards or not—the students wrote, were "apt to feel threatened" by a bricklaying robot. To union contractors, a robotic bricklayer would be an especially "tough sell." Summoning diplomatic skills, the future businessmen of America advised that union shops "have a greater means to resist the deployment of such a system."

The MBAs-to-be said there were twenty-five thousand masonry contractors across the U.S. but suggested the true potential customers were only the big boys: those that did more than $2.5 million of work a year. Then again, they also suggested that CR's first customer would be small.

The MBAs-to-be advised that a prototype—working not in a lab but in the real world, uncontrolled and nuanced—would be key to gaining acceptance. They said firms would want to *try before they buy*—and also said Construction Robotics would probably need a backup robot for these trials to go well. The industry, they wrote, would require a high "burden of proof" before embracing new technology. "The core value proposition," they concluded, was "strong."

On the other hand, they wrote, "the construction industry is not one with a strong history of adopting innovations."

Nothing like spending four grand to be equivocated to. In one way, what the business students wrote was so obvious it was worthless. In another, as things would show, it was priceless.

5.

In the Eyes of a Bricklayer

There's no such thing as a famous bricklayer. An unemployed bricklayer once wrote a bestseller, and briefly acquired fortune and prominence, but it faded fast. Where his father was buried alive in concrete, and his grandfather had been crushed to death in a tunnel, he drank with Hemingway, met the pope, and married a showgirl at a ceremony performed by Mayor La Guardia. All of this left him uprooted. "I became too sophisticated for bricklaying," he said, "and too confused to write." He had, however, captured the unique combination of humility and braggadocio that constitutes bricklayers. He died in 1992 and has been mostly forgotten. Schoolkids read John Steinbeck, not Pietro di Donato.

Even the *world's fastest bricklayer*, a title bestowed annually, isn't famous. Can you bring up his name?

Other bricklayers merely begat fame. The father of Yankees manager Joe Girardi was a bricklayer. So was the father of Yankees catcher Yogi Berra. Berra's father, who helped build the arena in St. Louis, did not like that his son got dirty for no better purpose than playing ball. This, of course, was a very Berra-ish opinion, since what was the point of the arena?

It was in the house of a master bricklayer that Thomas Jefferson wrote the Declaration of Independence, but does anyone remember the bricklayer's name? Another bricklayer is buried alongside some of the most famous Americans, including Paul Revere, John Hancock, and Samuel Adams. Born in 1692, he built Faneuil Hall, with bricks fired in Boston. When Faneuil Hall burned in 1761, only its brick walls remained standing. This mason's tomb reads simply: "Joshua Blanchard, a Mason."* Blanchard built the tomb himself. It's not far from Ben Franklin's parents; Crispus Attucks and the other four victims of the Boston Massacre; and even the French Protestants who came to Boston seeking refuge after Louis XIV revoked the Edict of Nantes in 1685—an act that drove many talented masons to North America.

If anything, bricklayers are infamous. A bricklayer started the Reichstag fire, and a bricklayer tried to assassinate FDR. One of the oldest mythical characters in the West, Renaud de Montauban, was a bricklayer. To atone for his sins, he'd been working for free on the cathedral in Cologne when, after a week, his fellow masons couldn't take it anymore. He was ruining their wages, so they hit him with a hammer and tossed him in the Rhine.

Some famous men dabbled with bricks, albeit temporarily. At Jones Beach, FDR slung mud for a few seconds. In Chicago, Enrico Fermi spent a couple weeks laying bricks made with uranium to create the world's first chain reaction. Woodrow Wilson's VP laid a gold-plated brick to complete the Indianapolis Motor Speedway. Lenin slung real mud, briefly. So did Winston Churchill; he wanted to build a cottage for his daughters. The press was captivated that the Chancellor of the Exchequer had taken up the trade, and in 1928 published a photo of him doing what Lenin had done. It's a great photo: Churchill is standing before a hip-high wall. He's addressing the inner two wythes of four. His trowel is in his right hand. His string line is straight. He's wearing a fedora, chomping on a cigar, and his shirt is spotless.

In response to the photo, one bricklayer wrote and told him his brickwork was "not on a par" with his state work. Another bricklayer, James Lane, the head of the local union, invited him to join.

"I do not feel that I am sufficiently qualified," Churchill wrote back.

Lane replied, "As time passes, you will improve your craftsmanship."

*The best memorial to a brick man is the one for Christopher Wren, who, rebuilding much of London after the Great Fire of 1666, relied greatly on bricks. His epitaph, in St. Paul's Cathedral, reads: "Underneath lies buried Christopher Wren, the builder of this church and city; who lived beyond the age of ninety years, not for himself, but for the public good. If you seek his memorial, look about you."

And then Churchill asked a question whose posing alone revealed his qualifications: Was there "a rule regulating the number of bricks which a man may lay a day?"

Lane responded: "Right honourable sir, there is no restriction whatsoever . . ."

Churchill joined and sent in five shillings. He got a membership card, listing his occupation as bricklayer.

This, not surprisingly, did not please union bricklayers. Churchill had fought a strike two years before and was not exactly a man of the people. One union bricklayer sent an anonymous letter, writing, "You damned old hypocrite. It would do you and the country good if you were forced to earn your daily bread by laying bricks instead of playing at it, and making yourself look like a fool." Still, one member offered to be Churchill's laborer.

And so, Churchill's union membership was discussed at the next executive council meeting. The most damaging claim against WC was that he had not said anything about how long he'd been working as a bricklayer. So the council revoked his membership.

Churchill refused to be ousted. In a statement, he pointed out that members of the union ought to "have assurances they cannot be turned out for political reasons."

The union did not reverse course. Churchill didn't let it keep him from laying.

And there was further proof that WC belonged: He had the attitude of a bricklayer. This attitude is best conveyed by the yiddish word *davka*, which means "stubborn in your own way." In union bricklayers, that attitude runs especially strong. In this regard, Scott Peters and Nate Podkaminer very much resembled bricklayers, even if, as events unfolded, bricklayers never thought so.

In any case, bricklayers are the kind of men who will dig deep to do something solely because they have been told they could not possibly do that thing. In short, they answer toughs with toughness. Consider two anecdotes from opposite sides of an ocean: In Washington, D.C., after authorities told a bricklayer working on the National Cathedral that they wouldn't seal his wife's remains next to Woodrow Wilson, he went ahead and mixed the ashes of his cremated wife into the mortar he was using. In southern Italy, so that taxmen couldn't confiscate any of his possessions, a bricklayer cemented his furniture to the floor and his pictures to the walls.

About their work, some bricklayers are humble and some are cocky, and many bear unique combinations of these traits. Christopher Wren, in 1671, approached three bricklayers on the same scaffold and asked what they were doing. One said, "I'm working." One said, "I'm building a wall." One said, "I'm building a cathedral to the Almighty." This false modesty today we call "humblebragging." Wherever a bricklayer falls on the modesty scale, he knows what George Meany, the first president of the AFL-CIO, called "the real feeling of satisfaction that comes to a building tradesman with the completion of a project built with his skill and labor."

It's tempting to attribute this attitude—stubborn, wily, tough, proud—to the existence of the bricklayers' union, but it's as likely that the formation and survival of the union was a result of the attitude, which runs in a bricklayer's blood.

The union was born in a turf conflict. A year before Appomattox, Baltimoreans organized the Bricklayers Union of the city of Baltimore, then almost immediately went on strike. The Baltimore and Ohio Railroad was rebuilding much of what had been destroyed in Confederate raids and, in need of bricklayers, was hiring recent army discharges. Union members objected. Upon tying up work, the union got the B&O to concede.

A labor newspaper in Philadelphia covered the strike, and the publisher suggested it was time for the various local bricklaying unions to form a national union. Organizations of bricklayers had survived the Civil War in New York, Providence, Pittsburgh, Boston, St. Louis, and Jersey City. Baltimore and Philadelphia liked the idea and agreed to meet. So on October 16, 1865, in a skinny brick building a few blocks from Independence Hall, nine men gathered. Five were from Baltimore, and four were from Philadelphia. They wrote a constitution, quickly adopted it, and, as if the writing project had left their syntax all jumbled, they called the new group the Bricklayers International

Union of the United States of North America. (Lost to history is why they didn't consult the typographers' union.) Testament to bricklayers' stubbornness and toughness, it is the oldest surviving national union in America.

During its first twenty years, the union stood up for a local striking here, a local striking there. Then, in 1886, when the National Builders Association declared itself an enemy of the union, the nature of engagements changed. The NBA, willing to negotiate only on wages and hours, sought to extinguish the union and establish, region by region, an "open shop," by which it meant a place where builders could hire any worker, union-affiliated or not. From Providence and Boston, the fight spread to Connecticut, then to D.C., Detroit, Louisville, Seattle, and Los Angeles. The fight was so costly that it left the union without momentum, and it stopped pushing for the eight-hour workday so that its members might have any days at all.

During the first two decades of the twentieth century, the union held its ground in this fight by tapping into its defense fund and underbidding open-shop contractors. By making and winning profitless bids, the union kept its members employed. Doing this was tricky: In Los Angeles, the union established its own construction company, and in El Paso, it established its own brick plant. There, the union also started a weekly paper, and either it was good reading or there was little else printed on paper, for it was gobbled up by schoolkids. Soon they began making fun of the chief anti-union contractor, so much so that his kids ran home in tears. Their scab father promptly wrote the union and agreed to hire union men.

All the while, union men checked the quality of the work done by strikebreakers. To the head of the union, open-shop work looked "unstable" and "hideous." But of course non-union brickwork looked bad to the union. What the union needed was for non-union brickwork to look bad in the eyes of the public—and good luck with that.

Over its long history, however, the union—bolstered by so much of a certain attitude—has remained perpetually adamant, which has brought it face-to-face with mighty foes. It has challenged behemoth construction firms, and U.S. Steel, and more than one chamber of commerce, and never flinched. In New York, union bricklayers once dug in so hard that the mayor intervened. Robert Moses, who liked brick, hated the bricklayers' union, and a generation later, so did most of big business. In 2010, the president of the bricklayers' union found himself handing out leaflets a block from the White House, because the administration of President Obama—*whom the union had endorsed*—had hired a non-union contractor for some restoration work. The union president, as it happened, was a member of the Democratic National Committee—which showed that friendship only goes so far. After the union made its case, the White House changed course. Had the union been around in the time of the Israelites, it would have taken on the pharaoh, and any day now, the union is likely to announce that its members intend to get Earth hours, Earth food, travel reimbursement, and astronomical wages when they're asked to build on Mars.

The union's demands could be confounding. During the Great Depression, when employment alone was heavenly, bricklayers demanded to get paid more than laborers. A decade later, the head of the union asked FDR to modify the design of the Pentagon, going with less reinforced concrete and more brick, thereby saving money and conserving steel. FDR passed on the request to the head of the army's construction branch—even though FDR once lobbied him against using brick. This exasperated the man, but he knew how to negotiate with bricklayers. Rather than meddle with the already poured walls, he proposed covering the building's interior walls with tile instead of plaster, giving plenty of work to bricklayers. When war broke out, though FDR had frozen wages, the union continually pushed for increases. Could it not be satisfied?

By then it could not, because for two generations, brickwork had been vanishing. To blame was not a cyclical economy—the panic of 1873 nearly wiped out the union, and the panic of 1893 left two thirds of the union's members unemployed—but cement, and steel, and elevators, and even railroads: pretty much everything modern.

At first, these innovations promised a reprieve from the chance of fiery death and wholesale property loss. American cities, composed predominantly of wood buildings, had been destroyed by massive conflagrations. In the 1870s, fires gutted Chicago, Boston, Baltimore, and Portland, Oregon. The Chicago fire burned six square miles and destroyed nearly twenty thousand buildings; tenements fell so fast and frequently that people said it sounded like an earthquake. The Baltimore fire burned for thirty hours and destroyed fifteen hundred buildings. In Portland, twenty blocks were destroyed. In the Atlantic Portland, a fire left ten thousand people homeless. Denver was leveled by fire. Nearly the entirety of Sacramento was lost to a fire. Downtown St. Louis was gutted by flames when a steamboat caught fire and, drifting past the waterfront, set it and three dozen more boats ablaze. New York suffered, too. The first time, during the British occupation, a third of the city was destroyed, and until it was rebuilt, people hunkered down under canvas sails tied to chimneys. Two years later, fire broke out again. New York's third major fire, in 1835, consumed seventeen blocks and destroyed six hundred buildings. It burned so hotly that it melted copper roofs, and it burned so brightly that it was visible in Philadelphia. In each city, builders turned away from timber to brick and stone, but they ran out of options as they built floors, which were necessary as immigrants poured into America. They ran out of money, too.

Where builders had tried encasing wooden framing in brick and tile, they quickly turned to elegantly tiled domes based on Catalan vaulting* and, in the Kahn system, hollow tile blocks arranged into arches and filled with concrete, with still more concrete on top. Concrete, builders quickly discovered, made possible big fireproof buildings, not to mention big profits.† Architects, moreover, were fascinated by concrete's potential. At the same time, steel frames and elevators made it possible to build ever upward and outward, quickly, with far less brick. And railroads could deliver huge quantities of sturdy building materials far faster than the local quarry. As cities like New York grew denser, fireproofing techniques once merely en vogue became

*See Carnegie Hall, Ellis Island's Registry Room, the Boston Public Library, the Smithsonian Museum of Natural History, the National Archives, the National Cathedral, Grace Cathedral, the Cathedral of St. John the Divine—or six hundred other libraries, hospitals, college halls, boathouses, state capitols, public baths, mausoleums, police stations, armories, synagogues, churches, apartments, hotels, and banks, scattered across thirty states, for fine examples of such work by the father and son named Guastavino.

†The wide availability of concrete was due to another modern technological novelty: dynamite.

enshrined in law. By 1891, a revolution in building methods had manifested: Where the Monadnock Building had load-bearing brick walls six feet thick, the Wainwright Building had only a veneer of bricks. It was a steel skeleton protected by a brick skin.

Therein lay the cruel technological fate that ever after afflicted bricklayers: As buildings got bigger and taller, roofers, electricians, plumbers, ironworkers, glaziers, carpenters, painters, and ventilation guys all gained work, even while the roofs, wires, pipes, beams, windows, and ducts they wrangled evolved. But masons—whose bricks and mortar remained nearly unchanged—lost out. Masonry went from fundamental—from *essential*—to just a veneer. By 1895, the amount of brickwork available had shrunk 80 percent. As Yogi Berra later put it: The future wasn't what it used to be.

In response to this sudden change, bricklayers at first held up their chins. Bricklayers in New York tried the Kahn system, hated it, and left it to laborers. They handed work to the American Brotherhood of Cement Workers, adamant that they would not work on buildings where concrete was used in place of masonry. They refused to touch concrete. At the same time, the union disparaged concrete-pouring and block-laying by unskilled labor as "the cheapening of construction," and—toward persuading architects of concrete's crappiness—started reporting all cracked and broken concrete construction. The union called concrete dangerous. The denunciation, of course, did not slow concrete down. By 1907, a concrete skyscraper had been built, and so many foundations were being made of poured concrete that stonecutters and stonemasons butted heads over the little stonework that remained. By 1920, reinforced concrete had allowed the construction of the biggest dome in history.

By then, union higher-ups recognized that their members had a flimsy grasp of the economy, and turned to protectionism. "We cannot sign away or give away to others work that is ours by right," the secretary wrote in 1901. "What we have we must keep, even if it involves a fight with kindred trades. We must take more—all cement systems that are replacing brick must be controlled by the international Union and done so quickly, for it is decidedly masons' work." Six years earlier, the temperamental Irishman heading the union had tried to make the same argument. He said it would behoove the union to "look ahead" and "try to protect ourselves against systems or patents which will drive us out of the market." For over a decade, though, an attitude stood in the way. The next union president called all the lost concrete work "one of the costliest and most serious mistakes in the history of the craft." The secretary agreed and instructed local unions to "demand jurisdiction over it"—if only by supervising concrete-pouring. "Some day," he wrote, "the bricklayers, who consider it beneath their dignity to control and do the cement work . . . will wake up and wish their dignity and false pride had not allowed the encroachment of cheap labor in this class of work."

Other trades, though, had little sympathy for the bricklayers, largely on account of their long-standing staunch independence. Forty years earlier, union bricklayers had refused to align with the Knights of Labor, and forty years before that, their fathers had resisted joining the Industrial Congress. They'd seen no point in going arm in arm with every protective society, benevolent society, and knitting group in the country for the sake of relations with industrialists. Not wanting two masters, union bricklayers resisted the appeals made by Samuel Gompers and the AFL for nearly thirty years. "We have no desire to, nor will we interfere in the labor troubles of other callings," the head of the union had put it in 1898. "We have all we can do to mind our own business without interfering in anyone else's." Having once stood up for plumbers in D.C., and marbleworkers in Philadelphia, the union grew wary of alliances and refused to strike for the sake of carpenters in Chicago. Bricklayers preferred not to team up with anybody—not with men who worked in shops, or with men who handled fabricated materials, or with secessionist stonemasons. When the Industrial Workers of the World emerged, union bricklayers found it terrifying. They didn't want revolution. They just wanted work. By 1924, the union's approach to survival had earned it such ill will that its head admitted, "We have been outside the labor movement so long that we scarcely have a friend among all the trades."

This situation left the union fighting not for wages but for survival. So it rankled when, as costs began to climb following World War I, anti-unionists pointed fingers at bricklayer wages that were rising while productivity stayed put. The insinuation, noted by Churchill, was that bricklayers were willfully limiting productivity. (To be fair, it took more work to lay a wall whose purpose was aesthetic rather than structural.) But so troubling was the allegation that, in 1928, the union declared that it "neither orders nor tolerates any restriction of work or output as to the number of brick laid in a day." Unofficially, bricklayers paid hourly knew they could make more money if they slowed down. Officially, the union vehemently denied the existence or support of such behavior. The union's stance may be found on the first page of the definitive book of union history, written by long-serving president Harry Bates. There, Bates decries any "restrictions in output"—calling such a notion a popular myth. In his words, any limit or attempt to limit output, in bylaw or otherwise, would meet with "instant disapproval" from higher-ups. Frown they may, but there was little they could do.

Amid the union's too-late conversion, internal squabbles over what little turf remained turned petty. Plasterers (absorbed by the union and added to its name in 1910), terrazzo workers, marbleworkers, cleaners, and pointers all stepped on one another's toes. Who exactly was justified in laying blocks? Who could claim cleaning and pointing old walls? Who was to lay mud for tile work, and to whom fell working with artificial stone? Regarding this last material, bricklayers and plasterers came to an agreement in 1913 that smooth stuff was for the former and fibrous stuff was for the latter. It was not long before plasterers figured out a trick. While building the New York Stock Exchange Building, they put burlap on the interior stuff and, over the objections of bricklayers, said it was theirs. Unsettled, this dispute rose to such a level that Elihu Root, FDR's secretary of state, was summoned to chair a tribunal. The way he divvied up the work was different: If it was made of gypsum, it was for plasterers. If it was made of cement, it was for bricklayers. Even this diplomacy was not sufficient, and sniping continued for years. In 1930, executive boards agreed that artificial stone, whatever it was, however fuzzy it was or wasn't, would be for bricklayers if it was over three quarters of an inch thick, and for plasterers if it was anything less. They thereby demonstrated that executive boards boast minds at least as nimble as those of six-year-olds.

Before that was settled, tile layers in New York said they'd tie up a church job unless bricklayers let them and only them set tiles on the outside of the building. The tiles were six by nine by one—nearly a thin brick. Bricklayers, given the choice between angering men or God, chose men, and conceded. The matter got the attention of the executive board, though, and henceforth the issue was settled: This time the cutoff was one inch. Unresolved was who would lay mud for the tile: tile workers or plasterers.

All the while, as use of traditional masonry fell, bricklaying slowly but unmistakably became a specialty. Until the 1920s, it had been common for general contractors to self-perform masonry, and for foremen and superintendents to be bricklayers. As brick use declined—and continued in that direction for the next century—placing bricks became a task left to subcontractors. This specialization would have great bearing on the reception of SAM by the construction industry, by mason contractors, and by the bricklayers' union.

The Great Depression was devastating to the building trades: With commercial construction halved and residential construction quartered, young men pursued other work. But World War II was rougher. The War Production Board froze construction. Bricklayers were out of luck until the war's end. And then, to the dismay of bricklayers, the postwar boom did not materialize as others had. Builders turned to prefabricated houses and to cinder blocks. Blocks and concrete now accounted for a third of all masonry, leaving the union no choice but to claim the work. It was dispiriting.*

A generation later, upheaval arrived as never before. It was precipitated by the Vietnam War, which had upset "normalcy" in the same way that World War I had: It sparked inflation. With so many young men an ocean away, domestic unemployment hit record lows. In 1969, only 3 percent of bricklayers were unemployed. With the short supply of labor in high demand, it could call for more than ever—and did. Construction workers demanded huge wage increases in the late

*Banning the closed shop via the Taft-Hartley Act (over the veto of President Truman), Congress salted the union's wound. Stinging further, federal law opened the way for state lawmakers to advance the open-shop fight by passing "right-to-work" laws.

1960s.* It wasn't highway robbery of the variety the Teamsters practiced, but the increases—from 6 percent to 60 percent—were more than other industries were getting, and it drew attention. Construction workers' strikes, which reached new heights during the Vietnam War, did, too.

The situation rankled Roger Blough, the CEO of U.S. Steel, an American behemoth that was having difficulty competing with foreign steelmakers. Blough pinned the rising cost of steel on rising construction costs. With an enemy in his sights, Blough began the Construction Users Anti-Inflation Round Table. He solicited big businesses, all wary of threats to the bottom line, and in a few years—after merging with the Fair Labor Law Study Group—had more than a hundred signed up. The group became the Business Roundtable and included Dow, Shell, U.S. Steel, Procter & Gamble, GE, Bechtel, Standard Oil, Union Carbide, Kennecott Copper, AT&T, Alcoa, Owens Corning, BFGoodrich, Sears, International Paper, and General Motors. These were mostly manufacturers and utilities that spent hundreds of millions of dollars annually on building factories and plants. Together with the U.S. Chamber of Commerce and the National Association of Manufacturers, they went after union hard hats: plumbers, mechanical contractors, electricians, operating engineers, carpenters, and masons.

The Business Roundtable sought to eviscerate construction unions and quash the power its workers yielded. Because it seemed that industry was beholden to unions, and unions were protected by a captive government, it wanted drastic changes. Its agenda included repealing the prevailing wage law, dismantling union halls, and curtailing union rights. Basically, it wanted to eliminate unions. It was the open-shop fight writ larger than ever, waged by the biggest guns in the country.

Toward this end, the Business Roundtable funded lawsuits that undercut union power, and mounted an effective anti-union publicity campaign. Blough called rising wages "the number one domestic problem of the country"—ranking the matter above the draft, Nixon, and civil rights. The Business Roundtable called construction workers

*The construction industry responded by manufacturing cranes, earthmovers, and lifts that were bigger than ever, obviating men. This, of course, prompted more than a few men to dream of bricklaying robots.

despotic and their wage increases "astronomical," "alarming," and "exorbitant." Of the same mind, the U.S. Chamber of Commerce called the situation consumer robbery and spread the message that union hard hats were rapacious featherbedders who didn't deserve to make as much as doctors or lawyers. Persuaded, *Reader's Digest* ran a few big anti-union articles, including one called "wage madness." *Business Week* said that unions were overreaching. *Engineering News-Record* called the situation "organized thievery." *Fortune* called building trades "the most powerful oligopoly in the American economy" and came up with the phrase "unskilled work, executive pay." Even the *New York Times* called construction unions "monopolistic" and arrogant, and said they "lived in a private world of their own."

Washington's men in suits did not fail to hear the group whom *Business Week* called "Business's Most Powerful Lobby in Washington." "I am damn sure," the labor secretary said, "that the construction industry is not entitled to the wage increases it has been getting." The chairman of the Federal Reserve Board called wage increases "unconscionably inflationary." The Nixon administration referred to the "wage-price spiral" as a "crisis," and Nixon himself called the industry "sick." When even the president of the United Auto Workers called the latest wage increases "excessive," how could they be wrong?*

As ever, though, the unions had a clever tactic at the ready. Seeing that big business was sidling up to Nixon, they figured they could saddle up even closer to the man. So in May 1970, tens of thousands of construction workers marched through New York City in support of Nixon and the war he was waging in Vietnam. To make it clear that

*They were wrong this way: A 10 percent wage increase for building a new factory amounted to "almost insignificant" increases in the cost of consumer products, as even the Business Roundtable's own analyst found. Which meant that where U.S. Steel blamed "skyrocketing" construction costs for its inability to compete with Japan, it should have examined its own failure to invest in the latest technology. The Business Roundtable also regularly pointed to hourly rates as a sign that fairness had floated away, but they forgot that almost nobody in construction worked fifty weeks a year. Many were lucky if they worked thirty-five. Construction workers were demanding more than they'd been making, but it wasn't like they were raking it in, certainly not like doctors or lawyers or the corporate executives besieging them. Last, as even Nixon recognized, construction workers were involved in an inflationary cycle, but they were its victims, not its cause.

they weren't hippies, construction workers also slugged a bunch of antiwar protesters and called New York's liberal mayor a "Commie rat" and a faggot—which made them look pretty far right. But it was just a show; they'd always been on the left. In fact, the marchers had been paid twice their normal wage to put on the show. It was a sneaky move, because Nixon at the time had no friends. For a while, the alliance—so rare for bricklayers—staved off the coming wave.

Then, in 1971, by proclamation, Nixon suspended the prevailing-wage law. A few days later, he announced a wage-stabilizing plan. The Business Roundtable called the move "politically courageous"—without mentioning its self-interest. An official from the U.S. Chamber of Commerce tried to put it in context. "If Ralph Nader and his co-workers . . . really want to protect consumers from exploitation, they could do no better than train their big guns on the wage monopoly in our nation's biggest industry." The irony, of course, was who had those big guns and who had the monopolies. A month later, Nixon established the Construction Industry Stabilization Committee, which limited wage increases to 6 percent.

By 1975, two million men had returned from Vietnam, and unemployment rose drastically. All of a sudden, there was a glut of construction workers, and unions became as protective as ever. In D.C. and New York, 90 percent of bricklayers were out of work. The lucky union member who found work took wage cuts of a third, and stories emerged of guys sabotaging their own work so they could do it again. They were, according to *The Economist*, "fallen aristocrats."

In fighting such a hard short game, building trades nearly abandoned the long game, which soon presented an even greater threat. By admitting few new members, practicing what came to be called "country club unionism," overindulgent local unions not only priced themselves out of industrial work, but they created a pool of non-union labor that underbid them on commercial and residential jobs. Locals lost grasp of their markets. Soon guys with union cards started hiding them because they were a liability. By 1979, *Fortune* saw the shifting tide as a reckoning. In a roundabout way, the Business Roundtable succeeded in weakening construction unions by fostering the expansion of non-union builders.

By the early 1980s, 90 percent of bricklayers in Alabama and Illinois were unemployed. Strikes were out of the question. And President Reagan, the former corporate face for GE, had as much regard for construction unions as the Business Roundtable.* Sticking it to them, he changed the way wages were set on federal jobs. These rates had been set according to a survey taken by the Department of Labor. Until Reagan changed things, the standard was that if 30 percent of those surveyed reported union rates, that became the wage. Reagan raised the threshold to 51 percent, which amounted to withholding meals from starving patients.

And so, between 1965 and 1995, membership in the bricklayers' union fell 30 percent. There are now a hundred and fifty thousand working bricklayers in America, and only a third of them are in the union. In response, the union consolidated its weakened locals. Where there were once hundreds, there are now thirty-five. Bricklayers, of course, resisted the change. The move consolidated power, but it put many bricklayers alongside strangers in strange lands—something they'd never liked. (In a conniving move, the union also changed its name to Bricklayers and Allied Craftworkers, which suggested, through vague broadness, strong bonds.) The union also established the International Masonry Institute, to formalize apprenticeship, training, and promotion. It's part R&D, part school, part radar. To an outsider, the IMI's creation may look like an offensive move, but it's all defense.

And still, in this new climate, bricklayers resisted change. In St. Louis, bricklayers opposed the use of story poles—because as surely as they made it easier to build corners, reliance on them also foretold of a future in which bricklayers lacked one more skill. That was the blood-born pride rising to the surface. Stinging further, a contractor was asking masons there to lay heavy blocks. The masons argued that the work of placing these heavy blocks should fall to two men instead

*Reagan's stance especially stung because it was a union hard hat—a carpenter—who probably saved Ronald Reagan's life. On March 30, 1981, when Reagan left the Hilton Hotel, where he'd been addressing the building and construction trades department of the AFL, Alfred Antonucci, the five-two, sixty-eight-year-old president of Cleveland's local 1750, was the first person to clobber John Hinkley Jr. on the head, causing his last shots to go wild.

of one. The contractor disagreed. This was in 2011. And so one thousand members of Local 1 struck . . . and in the fifty days before the dispute was resolved, the "how heavy is too heavy" argument—much like the "how thin is too thin" arguments of the past—was hashed out publicly, on talk radio. The way the head of the mason contractors' association saw it, every other mason in the country handled forty-pound blocks, so what was the problem? The way the head of the union saw it, anything over thirty pounds would injure masons, compromise their abilities, and affect productivity. The union prevailed and, as part of the two-men-to-a-big-block settlement, conceded on corner poles.

The bricklayers' union, today, is small. Since only four cents of every construction dollar go to masonry, the union pales beside electricians or plumbers. The BAC's peers are ironworkers, painters, sheet-metal workers. The BAC, though, tries to punch above its weight, as Jake McIntyre, the assistant to the treasurer, put it. That's how the BAC stands out. Among all the building-trades unions—which easily forge Republican ties—the BAC is categorically the most liberal. Union staff and leaders have been members of the Democratic National Congress (DNC) and gone on to chair the National Labor Relations Board (NLRB). Because the trade has suffered a long decline, and because masonry work remains as laborious as ever, the BAC is left of even the ironworkers' union.

Since the Business Roundtable first attacked, the bricklayers' union has found little appeal in pragmatists and centrists. Instead, it has rallied behind liberals who understand their plight. In 1980, the union abandoned Jimmy Carter and rallied behind Ted Kennedy. In 1984 the union supported Walter Mondale. In 1988 it supported Michael Dukakis. In 1992 it endorsed Tom Harkin, and in 1996, when Bill Clinton campaigned for reelection, the union didn't back him. In 2000 the union supported Al Gore, in 2004 the union supported Dick Gephardt, and in 2008 the union backed Hillary Clinton. The first time in a generation that the union backed a winner was 2012, when it endorsed Obama. In 2016, the union supported Hillary. As McIntyre put it, "Don't bet on our horse."

The union is still defending itself from the Business Roundtable, which has far more members than it had in the 1970s and is still pushing what it calls the effort to create, among workers, a level playing field. The union is also struggling in the aftermath of the 2008 recession. The only good news may be that the union has avoided Mafia interference. It seems likely that this stems from masonry's diminutive stature—a specialty trade with low margins—but I like to think bricklayers themselves—tough, resistant, never fans of alliances—staved off such entanglement.

Since the recession, membership has crept up, as have hours worked, but the average age of a union mason has, too. Today, the union figures a typical bricklayer is fifty-five years old, which means he started working in the climate of the early 1980s. If, like 10 percent of bricklayers, his father was a bricklayer, too, the younger bricklayer probably grew up hearing stories of the trade's heyday in the 1960s, and its demise.*

In one way, much of this history bolstered the case Scott Peters would eventually make: that the union should support his bricklaying robot. The current union president has taken to the radio in defense of alleviating the bodily toll induced by bricklaying. A former union president urged bricklayers not to let dignity and false pride steer them away from work. The union president who penned the organization's history said that the union's policy was "one of encouragement of any method that would make masonry work any easier or promote better structures." And the union's official stance—forged a century ago during the internal turf squabbles—is that it won't debate which tools bricklayers ought or oughtn't use.

In another way, though, it's easy to see how protectionist instincts and a tough mind-set, reinforced over generations, would spark resistance to SAM.

*Until baby boomers came of age, a third of bricklayers were sons of bricklayers. The boomers sent their kids to college and steered them away from the trade.

6.

Approaching Innovation

In the summer of 2011, Scott told Tim Lochner and Tom Coller that he wanted to pursue the bricklaying robot, since it seemed like a real business opportunity, and explained why he couldn't do it at GM. He asked, basically, if they would hire him—and because they found him intelligent, articulate, creative, and determined, they agreed. To the proposal, however, they added a condition: At PMD, Scott would not work entirely on SAM; he would also develop management software for the company. Scott's salary was a smidge lower than what he'd been making at GM, but it was close enough. Scott, now full-time in Victor, got closer to doing what he wanted.

Before the summer wound down, Scott did three more things: He filed a patent (Nate, Tom, and the RPI engineers' names accompanied his on the paperwork); met with the owners of Hydro-Mobile, who steered him to the annual industry trade show called World of Concrete; and interviewed a wild Lebanese man named Rocky.

He also read more start-up books: one about Google, one about Steve Jobs, one about GM's decline, and another called *The Lean Startup*. Of all the entrepreneurial books he eventually read, this was the most important.

Conventional wisdom, which spewed forth from places like GM and Kodak, was: *Thy products shall not be released until perfect, lest we disappoint customers and harm our reputation.* Eric Ries, the young Yale-educated author of *The Lean Startup*, argued for proceeding unconventionally.

Ries, whose first two companies had failed, proposed abandoning the quest toward the perfect thing; instead, he said, CEOs should aspire to build an MVP, or minimum viable product. A smart start-up, he wrote, ought to hurriedly ship a terrible, bug-ridden, unstable product—and then ask for customer feedback and change that product constantly. He said the path to success was not signed with a good business plan and paved with perseverance; but marked by a continuous process of experimentation, mistakes, and learning. (Perseverance he called a myth promulgated by magazine stories; to Ries, persevering was optimistically bumbling along in "the land of the living dead" while driving a company to failure. Far better than Caesar persevering was a listener behaving flexibly.) He argued that early adopters of any given technology, looking for advantage by taking risks, might have no interest in one feature but a great deal of interest in another—and what was the point in wasting six months in the hope that you'd picked the right feature? All that would do was precipitate a downward cycle: overworked engineers, frustrated managers, and then nitpicky conversations that drove planning to a halt, resulting in less building, less learning, and probably people quitting or getting fired. Any work beyond what was required to start learning, he wrote, was a waste. With an MVP, a start-up could immediately gain a "needed dose of reality" and ascertain what was required to get a customer to engage with a product and gloat to his friends about it.

"This is a hard truth for many entrepreneurs to accept," Ries wrote. "That world-changing product is polished, slick, and ready for prime time. It wins awards at trade shows and, most of all, is something you can proudly show Mom and Dad. An early, buggy, incomplete product feels like an unacceptable compromise. How many of us were raised with the expectation that we would put our best work forward?"

Because learning customer preferences outweighed whiteboard strategizing, and because idealizing an MVP challenged the traditional notion of quality, Ries said the path he was advocating felt dangerous. Indeed, proceeding down that path came with risks—of the legal and branding and morale varieties—and so Ries said that following the *Lean Startup* approach took courage. But he argued that the path wasn't as risky as it seemed. The fear that a big bad competitor might swoop in and steal a start-up's idea, he said, was largely imagined. A good start-up, adept at quickly building-measuring-learning, would always outdo the competition, regardless of what they knew, because the key to success was the ability to learn fast. A head start, he figured, was of negligible value; and in any case, a company couldn't operate in isolation forever, so it may as well give up the stealth-mode mentality and get busy learning how to learn fast.

Ries littered his book with the business jargon du jour—"pivoting" and "scale" and "disruption"—and a variety of catchy cherry-picked examples: Apple, Google, Amazon, Dropbox, Facebook, Ford, Groupon, Napster, Craigslist, Starbucks, Zappos. He even cited GM and Kodak.* What stood out, though, were his own oft-repeated phrases: "learn and iterate" and the "loop of build-measure-learn."

To Ries, being a small, obscure start-up facing immeasurable uncertainty was not terrifying but glorious. Such a situation allowed for under-the-radar experimentation and the chance to test business questions quickly. Marketing could be left for later—once a company's value was not the boss or the team but what they'd learned. These learnings he called a start-up's "essential unit of progress." Amassing these units, the way he saw it, was what led a company from brave start-up to humming innovation factory.

Scott fell in love with the idea immediately. He was a born experimenter and thought failing your way to success sounded way better than fooling yourself to demise. He could handle bumps, swim through ripples and waves and even swells. There'd be no avoiding mistakes, and rather than fear them, or Monday-morning-quarterbacking himself crazy, he thought learning and moving on sounded pretty good. As he put it later, "It doesn't matter if it fails. It matters if we learn. It's all about evolution."

The *Lean Startup* approach, though, was unmistakably if not expressly intended for start-ups selling software to consumers. Scott knew this; still, he figured he'd take the approach and use it at

*The learning-from-failure notion was not exactly revolutionary. Many years before Ries was born, Napoleon Hill rambled on about it, as did Winston Churchill, Ben Franklin, Goethe, Malcolm Forbes, and Henry Ford. Thomas Edison might have said it best: "I have not failed. I've just found ten thousand ways that won't work."

Construction Robotics, which was to make hardware for businesses. The differences were not negligible. Hardware is expensive, difficult to change, and very expensive to change quickly. Makers of consumer software might eventually have millions of customers; losing a few thousand early on posed little harm. And where consumers of software existed in huge quantities and could be easily be let go and forgotten—even apologized to—masonry businesses were far more rarefied and less easily abandoned. (In all of the United States, he knew, there were only twenty-five thousand mason-contractor firms!) If an MVP delivered to the construction industry was not clearly labeled as such—i.e., an early, unfinished, experimental prototype—word would get around fast that CR's machine was a piece of crap. Scott's gamble on the *Lean Startup* approach to innovation was certainly bold. It was also somewhere between ridiculous and insane.

Besides, the definition of an MVP was vague, endlessly open to interpretation. Viable to one guy wasn't necessarily viable to another. A man once sailed across the Atlantic on a sailboat the size of a toilet—but did that render his vessel viable?

For his first hire, Scott posted an ad on Craigslist. It mentioned electronics, controls, robots, and building. It was so enticing to Rocky that he thought it was a hoax. Scott felt the same way about Rocky: He was vibrant, more than just an electrical engineer. He'd inspected airport scanners and diagnosed blood analyzers, had built his own house, was even a pilot. One of the products he developed won an award,* but as important, he was tired of working around socially inept software engineers and delicately egoed PhD researchers. He wanted to *make* something.

So Scott stationed Rocky beside him, upstairs in a PMD outbuilding

*Rocky's laser recorder (which took two nonstop weeks to scan every frame of *Snow White*) allowed Disney to pump up the colors and remove the scratch marks in the 1939 reel. With such technology, Disney soon produced *Space Jam*, a feat of technical wizardry starring Bugs Bunny and Michael Jordan. A week before the Emmys, it earned Rocky an award from the Academy of Motion Picture Arts and Sciences, red carpet and all.

that was formerly rented to a seller of hardwood floors. First Rocky helped Scott reapply for the NSF grant, so Scott could pay him. Then he spent weeks debating with Scott about metrology systems, since he hated RPI's spinning-laser concept. To Rocky, that system was great theoretically, but totally worthless in the field, because it couldn't tolerate sun or dust and dirt. Plus, as far as a string replacement went, it was overly complicated. This assessment left Scott depressed. "We're building something here," he said. "It's great to say that's not okay, but what is?" Rocky said he didn't know—but figured he could come up with something.

Rocky and Scott argued about nearly everything. Rocky would present an idea, and Scott would disagree with him just to see how much he believed it. They butted heads over mechanical issues, which would set the framework for the culture at Construction Robotics. But the back-and-forth was never unwelcome; in fact, Scott encouraged his employee to speak his mind without sugarcoating. He'd ask forthrightly, *Why are we doing it this way?* He invited counterpoints, and—unlike Nate—never said it was his way or the highway. As he put it: "If you can convince me otherwise, I'll change my mind." When Scott wasn't convinced but wanted to acknowledge that he had at least considered a contravening viewpoint, the giveaway was his choice of words. "Noted," he'd say.

During these debates, the goal of three thousand bricks a day—3,000bpd—came up, even though the men had no idea how they'd ever get there.

Scott and Rocky didn't get the NSF grant, but one day Rocky went home, dug through his basement, and brought in some lasers of his own.

They applied a third time for NSF funding early in 2012. With RPI, they put a yellow Stäubli arm on an oscillating base—like a kid on Grandpa's knee—to test the robot's dynamic compensation, or its ability to steady itself under motion. Holding a marker and sheathed in a huge plastic bag, the condomed bouncing robot drew a house—a triangle on top of a square—to prove that it was up for handling bricks on a wobbly scaffold. Even in this elementary fashion, it was less than convincing.

When they got the NSF grant in the spring, Scott and Rocky earned a jolt of legitimacy and some freedom. Scott used both to hire a forthright and fastidious but mellow fifty-year-old named Kerry Lipp. Kerry had run his own residential construction company for twenty years, until, while reconnecting an electrical circuit, he fell off a ladder and broke his sternum, some ribs, and three vertebrae, as well as his hard hat. In search of a new trade, Kerry ended up in the school of instrumentation and control at Finger Lakes Community College. At Construction Robotics, Kerry turned an electric drill into a slump meter with a digital readout, and made a viscometer from a Sawzall and a spade bit. Essentially, he put together all the crazy things Scott dreamed up.

Meanwhile, Rocky used the newfound freedom to build a simpler metrology system. It used two (nonspinning) laser beams, fired from the right side of the wall. The first was a time-of-flight laser. Light travels at a known speed, so the time a beam took to get from the corner to a reflector on the gripper and back could simply tell a computer how far away the gripper was. This gave Rocky his X coordinate. The other laser, just above the first, would be set at a known height and distance from the wall being built, like an offset intangible string line. On the gripper, just above the reflective tab, a fiber-optic bundle behind a glass lens would catch this laser and give Rocky his Y and Z coordinates. With all three coordinates, the robot would know exactly where its hand was, very quickly. Two parallel lasers in a box and a lens and a reflector: That didn't seem so bad. Better still, Scott was convinced.

That summer, Scott hired another engineer, named Erwin Allman. In three decades at Kodak, Erwin had optimized the speed and efficiency of large automated machines—machines that cut rolls of film the size of a room, or made OLEDs, or deposited anti-counterfeit security markers. Scott stationed Erwin in the shop downstairs, five feet off the floor, on the steel platform of a Hydro-Mobile. He sat there beside Rocky, with whom he got along swimmingly. In fact, the two ex-Kodakers could be generally counted on to take the same side against their boss. Both men thought that Scott was failing to prioritize critical work ahead of premature demos. To keep things light, Erwin and Rocky avoided overloading themselves with technical mumbo jumbo.

Rocky called the nozzle a "smart sewer." Erwin called it "the anus." But the laser box, which had fifteen thousand dollars' worth of lasers and circuitry, got the best name. Rocky took to calling it "ten pounds of shit in a one-pound bag." Documents of no particular import or relevance he called "holy grails," just to test his colleague's astuteness. All the while, they faced the giant silver Stäubli.

Erwin's task at Construction Robotics was to make SAM work, or at least explain to Scott and Rocky why it didn't. In other words: controls. But Erwin, who had never programmed a robot, was stymied by motion compensation. SAM wobbled when it flexed its arm, and the scaffold had its own slow wobble, and the SAM-scaffold combo had a much more complicated wobble. Erwin sought to somehow make sure the end of the robot, using a measurement system, stayed still in space so that it could put a brick where a brick belonged. The problem had two sides: He had to know where the gripper was and then get it to dance—like a bird on a tree branch. Electromechanically, this was not simple. A hunch, in fact—based on a basic sensitivity analysis—told him that there was no getting "there" from "here." Erwin told Scott he didn't know how to solve the problem. That was when Scott hired Glenn White.

Because he'd tinkered with robots, Glenn was the first ex-colleague Scott thought of when he heard news of the fuel-cell lab's implosion. In college, Glenn had written a thesis on a robot of his own design, which had merit. Glenn wasn't able to read until fourth grade, but he could determine the gear ratios for the motor on his go-cart. He graduated first in his class in college and graduate school, the kind of student who took a test, went home, and redid the questions on his own time. He admired Feynman, Oppenheimer, and Einstein, while remaining in the dark about the definition or purpose of a blog. Since being saved by Christ, he'd given up rock climbing but maintained the wiry figure and slight hunch. Only a couple of years younger than Scott, he was endearingly earnest and hid wry humor behind sad eyes. He appreciated *The Simpsons* as much as Scott did, and his status as a believer rounded out the range of views among his colleagues, of whom Rocky was an atheist and Erwin was agnostic.

While he was at it, Scott hired another young ex-colleague to design the machine of his vision. Though he sported the goatee of a Harley rider, Tim Voorheis had the demeanor of a country doctor. A realist, he could be counted on to remain preternaturally calm and think his way through challenges. For Scott, Tim's challenge became building a 3-D schematic of the bricklaying robot, so that dozens more could someday be manufactured. Scott didn't even have one machine, and already he was thinking about persuading companies to buy so many bricklaying machines that he'd need sixty employees to keep things running.

As President Obama's second inauguration approached, Tom Coller told Scott that PMD would be erecting a new building. Scott suggested building the whole thing with the bricklaying robot he was designing. Tom responded with unmitigated logic: You can't build all of it, he said, but you can experiment on a little patch. Masonry was to begin in June, once the framework was up. That left Scott five months to build a prototype. It also left him and his employees in a drafty wood-lined trailer.

The owner of Hydro-Mobile had mentioned World of Concrete, the industry trade show, and that now seemed the perfect opportunity to hunt for components and get a sneak peak of the industry they intended to revolutionize. So Scott and Tim ran off to Las Vegas. What they saw at the convention center took their breath away.

Inside, amid forty-six miles of corridors, were concrete saws eight feet in diameter; paving machines thirty feet wide; jackhammers the size of rhinoceroses. There were pavement crushers, curb setters, cranes and lifts and scaffolds of a dozen varieties. There were not just concrete mixers but concrete *factories*, half the size of a battleship and not much cheaper. At the second-biggest show in Las Vegas, nearly fifty thousand attendees in everything from baseball hats to business suits wandered around, checking out the latest machinery on display from thousands of companies. The people and machines seemed nothing so much as steroidal: giant hydraulics, enormous necks, acres of gleaming powder-coated steel. While holding their jaws (and much smaller jowls) shut, Scott and Tim hunted humbly for mortar pumps and admired Leica's expensive laser tracker, its self-tracking eye no less awesome than the red all-seeing eye of Sauron.

Outside, in massive parking lots, there was more. You could drive a dump truck up huge piles of gravel, or ride a terrazzo polisher as if it were a bumper car, or play soccer with a bulldozer, or watch the Bricklayer 500—where two dozen men laid bricks as fast as they could for one mad hour, competing for the title of World's Fastest Bricklayer. The event was staged between bleachers, along whose sidelines rabid, costumed, sign-holding fans cheered and drank beer and ate hot dogs under the bright Nevada sun. As the event unfolded, it was narrated by an emcee and shown on a thirty-foot Jumbotron in the corner. Scott and Tim grabbed seats in the bleachers and watched as a man from Creative Masonry in Limestone, Tennessee, placed 634 bricks in one hour, for which he took home five thousand dollars and a bright red Ford truck.

Taking all of it in, Scott turned to Tim and said, "Wouldn't it be cool if we were here?"

7.

The World's Fastest

In the early 1800s, a bricklayer earned around three dollars for laying one thousand bricks, then an extra dollar for every thousand bricks placed above the second floor. As he went higher, the work got harder—more complicated and riskier.* Placing bricks in the Flemish bond (alternate headers and stretchers) garnered an extra quarter. On short buildings, one thousand bricks, between dawn and dusk, was a day's work. On buildings four or five stories tall, a bricklayer could count on laying two thousand bricks a day, because he could basically throw bricks at walls three feet thick. The brickwork didn't have to be pretty except on the outside. Of course, a bricklayer also had to pay for his laborer, who charged one dollar for every thousand bricks.

Then the industrial revolution came along, and around the turn of the century the new choreography of construction discombobulated everyone's notion of productivity. The National Association of Manufacturers estimated that in just a decade, the productivity of bricklayers fell 70 percent. The Royal Commission on Labour found the same thing. It interviewed a mason who said, "Men do a great deal less work now than they ever did. Where it used to be the custom for a good bricklayer to lay a thousand bricks a day, three hundred or four

*The same holds true today: On buildings below four stories, getting tubs of mortar and cubes of bricks onto a scaffold requires a machine like the Gradall, on whose end Nate once dreamed of sticking a Stäubli. On buildings four stories and above, getting supplies up to masons requires a mobile crane . . . and on buildings so high that even a mobile crane won't cut it, construction elevators or tower cranes become mandatory.

hundred is about the usual thing now."* That's been the complaint for the last 125 years.

In that new climate, fast bricklayers stood out. One notable standout was Frank Gilbreth, who became known at age thirty as Boston's fastest builder, having completed MIT's Lowell Laboratory, with a thousand pilings and a million bricks, in ten weeks. The performance earned him authority. Gilbreth, who became a follower of motion-studies guru Frederick Taylor, was interested in pinning down the optimal way to lay bricks. Though he wasn't a trained engineer, he devised his own scaffold, so men wouldn't have to stoop to slop up mortar and grab bricks. He figured humans ought to behave like efficient machines. Toward this end, he sought to synthesize all that was known about the art of bricklaying, so that knowledge handed down by word of mouth, from all the old countries, would not be lost, and superior techniques could be promoted. This synthesis became a book published in 1909. In it, his aim is clear. "The bricklayer must increase his output," Gilbreth wrote. "He must remove all obstacles that make for reduced output. He must use every device that will lessen the cost of brickwork." In line with the construction firms around him, he sought to promulgate the speediest bricklaying possible, while making allowances for the psychology of bricklayers, which he'd come to understand.

This psychology was delicate. "Some bricklayers with good intentions," Gilbreth wrote, "cannot be made to leave off their old habits . . . it is not wise to interfere with this type of man." On the other hand, there were "first class men," who ordinarily worked two or three times faster than others, most of whom put effort into disguising their slow pace. This led to some moralizing. "The work of a bricklayer is generally indicative of his character," Gilbreth wrote. "If he is dishonest he will do dishonest work and cover it up before the flaw is seen." Fast or slow, honest or deceptive, set in his ways or not, a bricklayer was universally likely to abide by one more principle. "Bricklayers," Gilbreth wrote, "are not proud of, nor interested in, a piece of work, unless they build it all."

*There were exceptions. In 1901, James Stewart compelled hundreds of bricklayers working for the British Westinghouse Electric Company to lay upward of two thousand bricks a day. For a month in 1905, bricklayers building a ten-foot tunnel for the Detroit waterworks averaged thirty-six hundred bricks a day.

That said, not being an engineer, Gilbreth had enough insight to know competition between men was the best way to get big numbers. It was only human nature, and such competition already had a history.

In the fall of 1870, a Philadelphian named George Talleman bet a hundred dollars that his man, Wade Cozzens, could lay five hundred bricks in twelve minutes. A Mr. Farley took him up on the wager, and on the afternoon of Friday, November 4, two miles south of Independence Hall, the contest was settled. According to newspaper reports, Cozzens laid ten bricks in the first second and twelve bricks in the tenth second of the third minute—though they don't say whether Cozzens laid bricks with or without mortar. He had three men supplying him with bricks, and their six arms could not keep up with his two. By the time twelve minutes were up, Cozzens had laid 702 bricks—for an average of 59 a minute, or about one a second. To capture that kind of speed on film, you'd have needed Eadweard Muybridge.

Seventeen years later, on October 12, a twenty-four-year-old Chicagoan named Frank Stoewahs laid 162 bricks in two and a half minutes, for an average of 65 a minute, or just over one a second.

Twenty-eight years after that, ten miles west of Manhattan, competitive wall-building was elevated to new heights. At the ball grounds in Athenia, New Jersey, "rushfest" was held, a one-hour spectacle involving seven teams of two hailing from seven different states. All built identical walls, forty-one feet wide, eight inches thick, and six feet high, and did so under the watchful eyes of two referees.

The contest began at a quarter after three p.m. Flutestein and Allairdair, the team from New York, laid enough mortar for a dozen bricks at a time, using a "long sweeping swing." Dromia and Ryan, from New Hampshire, kept pace with the New Yorkers. O'Sullivan and Always, the duo from Connecticut, used the "Boston Dip" and showed impeccable rhythm. Michigan's "Eat-em-up" Jack and "The Bug" were also fast but fell seven bricks behind. Neville and "Old Pop," of Boston, were disqualified for raising their line two at a time. The local boys, the "Two Toms" of Passaic, especially impressed. A newspaper said they were "sleek as antelopes" and revealed the "graceful undulating movements of the swan." According to that same newspaper, the event was more exciting than horse racing, bicycle racing, or baseball.* As such, it called the Two Toms "splendid specimens of manhood."

As the final minutes wound down, only a few bricks separated the leaders. The Two Toms, apparently, were "veritable demons in a finish" and helped produce the "most exciting event ever witnessed in this part of the country." As they laid their final bricks, the roar of spectators was so loud it echoed off Dundee Dam and made it across the Ramapo range. Where most walls were uneven and full of voids, the wall built by the Two Toms was made of bricks entirely level, plumb, and true, with bed and head joints uniform and beautiful. More important, the hometown boys finished three bricks ahead of the competition. In one hour, the pair placed 1,920 bricks (or 960 each), for an average of 35 a minute.

One-hour competitions became the new thing, particularly solo endeavors, especially in England. In late November 1924, in Treeton, Yorkshire, Christopher Hull laid 809 bricks. A month later, he laid 844. The following day, at Scarborough, John Wood laid 879 and stole the record.

During the middle of the twentieth century, by which time Frank Gilbreth was gone, record brick counts went up only on the big screen. Andrzej Wajda made a movie about a Russian bricklayer who laid *thirty thousand* bricks in one shift. That was a brick every one and a half seconds for twelve straight hours. In a different movie, an Italian bricklayer decided to build a house in one night, because he knew the police couldn't evict anyone from a structure with a door and a roof.

And then, on February 21, 1987, Bob Boll showed up at the U.S. Brick Olympics. The event was held in the Midwest, and the goal was to see if an American could break the Guinness World Record for one hour of bricklaying, which had been set by an Englishman in 1984. Curiously, the English record was under eight hundred bricks—fewer than Hull and Wood had laid sixty years earlier.

Bob Boll had been laying bricks since 1972. He was six-two,

* Which was saying a lot, because Ty Cobb was hitting every other ball and stealing four bases a week.

thirty-three years old, the father of five, and had just become a devoted Christian. His great-grandfather had been a bricklayer. Above his mustache, Boll wore wire-rimmed glasses and a blue headband. The competition was mostly union bricklayers from Michigan, Chicago, New York, and Minnesota. All built double-wythe thirty-foot walls as fast as they could. Four men broke the Guinness record—but Boll broke them. He placed 914 bricks, all of them level and plumb within a quarter inch. Anything less and he'd have been disqualified. Boll had been so busy laying bricks for work that he didn't have any time to practice; he hadn't built a double-wythe wall in five years. He did, however, lose twenty pounds by doing sit-ups and push-ups. For his effort, the man who used to hate it when people watched him work won three grand, a new Dodge truck on whose side a decal proclaimed BRICKLAYING CHAMPION 914 BRICKS IN 1 HOUR, and a plaque from Guinness. The VP of a company that made levels wrote him a congratulatory letter and signed off with "Good luck and keep that mud flying."

Boll kept the mud flying, but not at the U.S. Brick Olympics. The event collapsed after sponsors bailed. For fifteen years, Boll's title went unchallenged. Then, in 2003, the event was reincarnated as the Bricklayer 500. The first BL500 was held in Las Vegas, and only seven bricklayers competed. Boll didn't even hear about it. Wayne Phipps, an Arkansan son of an Arkansan bricklayer, won by laying a paltry 539 bricks.

The next year, Boll still had his 1980s mustache and was still, in his own words, "kickin' the brick in." This time, the event was held in a parking lot beside Orlando's convention center. Boll brought his family, including his son Paul, who worked as tender and kept his father well stocked in bricks and mortar. With 765 bricks, Boll won the world championship title again and this time took home ten grand.

The Bricklayer 500 returned to Las Vegas in 2005, and there the mortar tripped Boll up. It was not creamy, like in Florida, but chunky. For whatever reason, Vegas also made Boll so nervous that he flubbed his performance; his younger brother Mike beat him by six bricks.

The next year, Mike Boll got second place, and the year after that, he won a second time, just like his brother. And then, in 2008, a North Carolinian named Garrett Hood beat Mike Boll by eleven bricks.

Garrett Hood had first taken a bricklaying course in tenth grade. He started competing the next year and got first or second in a dozen local events. His dad, a general contractor, called him a "bricklaying machine." He also said, "If you slow him down, you mess him up." Hood played baseball in high school, on the mound. The only pitch he had was a fastball, up around 86 mph. Fast was his thing. He got so into bricklaying that when he was accepted to Tennessee State's engineering program, he didn't want to go, because he didn't want to stop slingin' mud just to look at books.

He married into one of North Carolina's big residential masonry firms and became a supervisor. He kept laying brick and often practiced laying quickly while his dad timed him. He had to buy two trowels a year because he wore them down until they were so small they didn't hold enough mortar and the sides were curved, rendering them useless. With each brand-new trowel, he suffered through the first month of use and the factory-made roughness that kept mortar from sliding off easily. After four weeks, mortar polished it like glass, allowing him to sling like a demon. His firm was non-union, and like the other big North Carolina firms, it offered bonuses to bricklayers who exceeded a minimum number of daily bricks, compelling them to hustle the minute they got out of their trucks in the morning.

Hood came in second to his coworker in 2009 and won the Bricklayer 500 a second time in 2010. That year, he placed 1,011 bricks, setting a new record. He traded in the two trucks he'd won for a brand-new Ford F350 longbed with all the bells and whistles.

As Hood's mortar was setting, two New Jersey masons had a few drinks and then came out to inspect the winning walls, which stood like wide tombstones in the abandoned parking lot. *This is the winner?* one said to the other. *This looks like shit. How can this be the best bricklayer in the world?* (It didn't help that the "world's fastest bricklayer" was not, and had never been, a New Yorker.) A Spec Mix guy overheard the men, and that was when Spec Mix, the event's sponsor, changed the rules.

Hood's huge brick counts came at the expense of impeccability, and Spec Mix figured that while it was one thing for the company to promote speed, it was another to promote shoddiness. So they tightened the competition rules nearly back to the standards of the U.S. Brick Olympics. Henceforth, competition walls had to be not only level and plumb, but every joint had to be cut flush and set within tight tolerances, and bricks couldn't be lipped (protruding) or tipped (slanted) or chipped or—heaven forbid—backward. Assessed by judges, all infractions would incur penalties. A backward "shiner" would earn an embarrassing deduction of fifty points. A tipped brick would earn a deduction of twenty-five points. A wall with more than twenty voids would earn a deduction of a hundred points. Consequently, in the ten minutes of cleanup time allotted to competitors after their sixty minutes of furious bricklaying were up, competitors took great care to plumb up and level and strike and even brush their walls. (They were forbidden to lay more bricks in this time.) About the only thing they didn't do was give their walls a shampoo. In any case, in the four years after Hood's 2010 dash, no mason topped seven hundred bricks, because competitors dialed back their speed so they could rev up their quality.

In 1890, while Frank Gilbreth was honing his speed as a bricklayer, another brick man fifty years his superior was standing up for brickwork's quality. This man, an Ohioan named J. W. Crary, was the inventor of a brickmaking machine and son of a brickyard owner, and generally of the opinion that a brickmaker could never go bankrupt.* He loved brick and was lucky to live before its decline. "The highest possible proof that brick is the most superior building material in the world," he wrote, "is the fact that . . . the entire brick work of the whole world is imperfect." It was a strange kind of praise, but praise it was. He could have observed that without brick, there would have been no great waterworks, no chimneys, and no mighty furnaces for making steel. But Crary had a subtler point. Brick, he wrote—whether it composed a warehouse or a church, a grimy factory or a splendid manor—had "to stand silent and defenceless, and bear the sins of botchers and the aspersions of novices." In its capacity to do this, he figured, brick was unrivaled.

*Go figure that Warren Buffett owns America's largest brick company.

And Crary was right: That a single building material produced in astounding abundance and variety could successfully mask so many sins said a lot about the art of bricklaying. A century and a quarter later, it also says a lot about the poor eye of the casual architectural observer, because those sins of imperfection surround us, and you don't have to be a judge at the Bricklayer 500 to spot them.

Believe it or not, a good wall is hard to find. Examine a brick wall closely, and it's easy to find something out of whack. Most obviously, the bed joints on different courses may be of differing thicknesses—on account of the masons who laid them, or the laborer who mixed the mud, or the weather, or the company that made the bricks, which vary in size.* Perhaps, between a section laid one day and another the next, there's a jump in bed thickness, or from mortar one shade of gray to mortar of a slightly different shade of gray. Or maybe the beds on a course or two reveal a gentle smile, a just detectable sag, the result of masons laying too much at a time. Or, between a section laid in the morning and a section laid in the afternoon, the profile of the joint, easily probed with an index finger, feels different. Maybe some bricks protrude out of plane: Noon, when the sun is high, is a great time to check. So are the first few minutes of a calm rain. More likely, the head joints don't line up course after course after course. Stand across the street and hold up the straight edge of a book or magazine to check. Better yet, stand close and look skyward. Do the head joints produce straight dashed lines? What about the wall's expansion joints? Because masonry expands and contracts like any other material, brick walls contain continuous narrow joints every twenty-five feet or less, filled in not with rigid mortar but with a compressible sealant. Are these straight and plumb? The ultimate test: Put your head against the wall

*Early in the twentieth century, it was fashionable to build walls with very thin beds of mortar. In some places, fat beds were the rage. Where I live in Colorado (I've never seen such work anywhere else), it was apparently A-OK not to tool the joints and just leave them smooshed and dripping. Every time I see such masonry—which I want to call Unstruck or Strikeless—I think of Dalí's clocks, oozing downward. What was no doubt attractive about this style of masonry is that nobody could tell if the work was shoddy or not. Examining it is as futile as looking for brushstrokes in a Pollock painting. The same goes for an even more outlandish style called "drunk bricks." I've seen it only in Denver.

and look diagonally upward—do the staggered head joints converge, like railroad tracks in perspective, to a distant point?

Bad brickwork, of course, is relative. In ancient Babylon, brick houses regularly fell down and killed their occupants, because sun-baked bricks, reinforced with straw, could handle only thirteen pounds per square inch. (Regardless, the builder was held accountable and killed.) Kiln-baked bricks, which could handle a thousand times more force, were too expensive for everyone except the pharaoh, who used such bricks in towers and tombs. In the modern era, bad bricks—known as "salmon" or "bats"—could allow for a jailbreak. Most of the time, though, bricks were exceptionally durable, and could be counted on to reveal the care (or lack thereof) with which they'd been laid for centuries, and even survive disasters. Some proof: Bricks from a church destroyed in the Great Chicago Fire were reused at Lake Forest College. Bricks that fell in San Francisco's 1906 earthquake are now in Seattle's Pike Place Market. Ever the brick evangelist, J. W. Crary framed society's relationship with bricks thus: "A great, wise, healthy, good people must have good buildings, public and private. It is impossible to have them without good brick, and good brick cannot be sufficiently produced without intelligent skill, and intelligent skill cannot be sufficiently grown and qualified without literature, and literature cannot be had and made available without talent and fitness in publishers . . ."

While Crary's promotion may have veered toward the exaggerated, it was warranted, because bricks, as individual objects, barely registered then and register now even less. Few objects—not spoons, not pencils, not even paper clips—are admired less than bricks. Bricks seem so unremarkable that in all of the United States, only one rickety museum is dedicated to their preservation, and the active members of the International Brick Collectors Association number only a few hundred. In other words, the probability of meeting a brick collector is lower than meeting someone who was struck by lightning.

To the "brickers" of the IBCA, what matters is neither the speed with which a brick was laid, nor the quality by which it was placed among its brethren, but history. They are firm believers in the saying "A brick without age is like a kiss without squeeze." In their minds, the word "brick" does not metaphorically represent a measure of stupidity ("dumb as a brick") but means, as it did 150 years ago, a "facetious, funny fellow who says smart things." Bricks, to them, are incredible. Though steel gets all the credit as a modern marvel, bricks, in their eyes, deserve just as much. Brickmaking requires mastery of fire, of chemistry, of labor, of conflict, and of time. Bricks reveal just how dark the Dark Ages were: so dark that, throughout Europe, no bricks were fired. Someone itching to build a chimney or well had to go scavenge an old Roman brick.

As such, while IBCA members generally swap bricks for free, they pay up to $250 for a good collectible brick. Such a brick could be from the 1893 Columbian Exposition, or from Thomas Edison's New Jersey lab, or from Abu Ghraib, or from pre-statehood Oklahoma, as indicated by the letters O.T., for Oklahoma Territory, stamped on one side. Of interest are bricks from the Hungarian People's Republic, with a hammer and sickle stamped on one face, and bricks with Saddam Hussein's image. Such markings have a long history: Nebuchadnezzar stamped bricks with his own name; so did Ramses. Other ancients marked bricks with their fingers or with royal hieroglyphs. Brick collectors get excited about DURANT bricks, which are rare, or any brand of bricks made in South Dakota, which are even rarer. One of the most famous bricks in the country, from Kansas, has an engraving on it that says DON'T SPIT ON SIDEWALK—and was put down early in the twentieth century by a bricklayer helping Dr. Samuel J. Crumbine, Topeka's public health officer, who sought to limit the spread of tuberculosis. For about twenty-five bucks, you can buy one of Crumbine's nudgy bricks on eBay.

The IBCA was formed thirty years ago as an offshoot of a group of barbed-wire collectors whose spending habits, apparently, were shocking: They spent thousands of dollars on stubby little pieces of rusty wire. Collectible bricks, by and large, are far more affordable—such that members fill their basements with collected bricks, cover their kitchen floors with collected bricks, pave their front paths with collected bricks. One woman in Wyoming has five thousand bricks in her (uninsured) collection, and one man in Oklahoma has fifteen thousand bricks in a building of their own. Another Oklahoman was so captivated by bricks that he spent a decade researching *Made Out-a Mud*, a book that limits itself solely to bricks from the state.

But why limit yourself to one state? Independence Hall was made out of bricks. Monticello was made of bricks. The oldest building at the oldest college in North America is made of brick. Ellis Island's Registry Room—the first American building into which millions of immigrants stepped—is clad in brick. The batting room at the old Yankee Stadium—a place occupied by a frenzy of American heroes—was clad in brick. Under its limestone, the Empire State Building contains ten million bricks. America, by and large, is made of bricks!

Even Rochester is famous for bricks. Rather, make that infamous. The campus of the Rochester Institute of Technology, composed almost entirely of fifteen million Belden bricks, is kindly called "brick city," and less kindly considered one of the country's biggest brick blunders. The buildings, most built in the 1960s in a twist on brutalism, look fantastically . . . uniform. *Travel + Leisure*, citing the "ceaseless" Belden iron spots, named the campus one of the ugliest in the country—paying no heed to the speed or quality with which they were laid. During the last decade, the university has sought "warmer materials"—precisely the kind of thinking that Scott Peters would be up against.

8.

Mortar Mike

As tends to happen in construction, the schedule for the construction of PMD's new building got pushed, and luckily so, for Scott was still not ready. To help with non-engineering matters, Scott hired Zak Podkaminer, his brother-in-law. Zak had worked construction (on jobsites and in his father's office), studied management, and, for years, worked in Boston as an operations consultant for Fidelity. Like Scott and Nate (and Frank Gilbreth), he was an advocate of efficiency. Officially, Scott put him in charge of marketing and IT. Zak created the company logo, came up with the naming scheme for SAM (technically, the machine's name was SAM-100), and set up everyone's email accounts. Unofficially, he became CR's business manager, doing research, planning, financial modeling, business development, grant writing, part ordering, and price negotiation. He was a skilled negotiator—an inveterate bargainer, really*—and a good salesman. He might have been Nate's most troublesome child, but he'd long encouraged him to pursue the bricklaying robot. He also got along great with Scott. So, even less officially, he became Scott's helper. When brick deliveries arrived (Belden Brick, ever eager after the recession to promote brickwork and reverse the tide toward glass and concrete, donated ten thousand bricks to Construction Robotics), Zak ran out and manned the forklift, and when it snowed, he plowed the loading dock. This position also left

*At Home Depot, he managed to get the military discount by pointing at his close-cropped hair.

him in charge of marketing, a task he approached with far more energy than skill. Then again, Eric Ries had advised disregarding marketing, so maybe it was of little consequence.

That still left engineering matters. The engineers around Scott remained mystified by mortar. From trials that involved making small brick sandwiches—bricks were the buns and mortar was the meat, some raw, some well done—they learned how the meat squished and figured out the consistency they wanted. They knew the desired dosage—enough to produce a bed of three eighths of an inch—and, from trials, learned how to deliver it. They'd tried a peristaltic pump (if you step on a hose, water comes out), a positive piston pump (a big syringe), and finally, a progressive cavity pump (a big screw in a tube). To test this last one, Rocky brought in a garden tool that drilled holes for flower bulbs. He put it in a three-inch plastic pipe, and it seemed to work. The engineers knew that on each brick, they had to apply more mortar in the middle and less on the edges, since the placement of a brick forces some of the mortar out. But application stymied them. They couldn't get mortar to reliably stick. On any given try, after the robot brought a brick to their nozzle and moved from bed to head, maybe half would fall off.

What looked like evolution to Scott often looked like chaos to the engineers working for him. *Forget that thing!* an engineer would yell. *We need a whole new thing!* As Tim Voorheis later put it, "If you have no clue how to solve a fucking problem, you just try anything." Half the time, Tim thought: *This is never gonna work.* Two other men had come and gone after only months. One had spent his life at Kodak, and one had just earned a PhD. Both were too slow and methodical for the working style at CR. Even Rocky, who enjoyed the experimentation, was disheartened by Scott's schedule; they kept moving forward before fully understanding the interaction of systems. They made guesses and left specs loose. An auto manufacturer working like that would be lucky if the wheels didn't fall off. Kerry later called it Hail Mary engineering.

Then Mike Oklevitch started and quickly revealed that the guys were not thinking about mortar the right way. As a chemical engineer at Kodak, Mike Oklevitch had spent twenty years working with slurries, dispersions, and extruded coatings on films, so even though he'd never touched mortar, it clicked immediately. Scott had found Mike Oklevitch in the same program that Kerry had attended, and hired him as an intern. He was fifty-two years old and fully gray, which earned him, alone among CR's engineers, full-name treatment—at least for a while. Very quickly, he earned the nickname Mortar Mike.

Mortar is strange stuff. It's a non-Newtonian fluid, meaning its viscosity does not change only on account of temperature. (Nearly three times as dense as water, it weighs about 150 pounds per cubic foot.) It's not thixotropic, like quicksand; nor does it break down and get thinner (a phenomenon known as rheomaiaxis), like shampoo or ketchup. It's rheopectic, which is much rarer. More specifically, it's dilatant. It thickens.

Like bread dough, mortar is a matrix, whose predominant nonwater ingredient is sand. Three quarters of the dry stuff is sand. The rest is cement and lime, with maybe a dash of modern additives. The strongest mortar has lots of cement, little lime. The weakest has lots of lime, little cement. Over generations, this last stuff will leak and need repointing (especially if the wall traps moisture inside via a coat of paint). To make things easy—because you don't want mortar stronger than the masonry units to which it's bound—ASTM has categorized the five most common types of mortar, using every other letter from the word "MASONWORK." So, type M, capable of withstanding 2500psi, is the strongest, and type K, capable of withstanding 97 percent less, is the weakest.

When the ASTM came around, two types of cement were competing for the attention of masons. There was hydraulic cement, produced in Fayetteville, New York, since 1819, and there was Portland cement, a variety patented by a British mason in 1824, even though Brits had been tinkering with the recipe for a while. Portland cement had been named after the stone found on the Isle of Portland, in Dorset, because it bore a resemblance. Portland cement didn't set as fast as the hydraulic stuff (which was useful), and it hardened faster and was stronger. Making it entailed heating limestone and clay to get clinker, but it also entailed adding gypsum. Portland cement, though, wasn't manufactured in the U.S. until 1872, three years after men laid a rail line across the continent and rendered the Erie Canal—born by so much hydraulic cement—obsolete.

And still, for a generation, masons found hydraulic cement good enough. In 1899, nine tenths of the cement produced annually in America remained the natural/hydraulic stuff. That was over three billion pounds. By 1903, as buildings taller than ever were going up faster than ever, the split was 50/50. And by 1910, the situation had flipped. As the nation prospered and grew, masons ditched the lime, resorted to Portland, and built upward. By 1970, there was not a manufacturer of natural cement left in the country.*

Unless you're Mike Oklevitch, mud today includes Portland cement. Mike Oklevitch prefers generic mortar, free of proprietary additives, and also enjoys "play mortar," free of cement. Without cement, Mike Oklevitch freed himself of the worry that mud in SAM would harden up, and eventually eliminated concerns regarding test walls that were so heavy they were unliftable. After building walls, he could disassemble them, scrape the mud from each brick, and reuse both the bricks and the mud. But first, play mortar, free of cement, allowed Mike Oklevitch to test adhesion at a comfortable pace.

Before Mortar Mike showed up, the engineers had believed that the magic of getting mortar to stick to a brick was all about consistency, and they assumed consistency was all about moisture content. As mud thickens, masons add water and mix. It's a chemical instinct, like adding milk to the pancake batter at the bottom of a bowl. Because the moisture content of a mortar mix from Chicago was not the same as one from Chihuahua, the engineers had even come up with a moisture ratio equivalent and figured out the threshold below which the stuff was so thick they couldn't pump it. They knew that getting it to stick required pressure and application at an angle, so as not to entrap air. But mortar is not like pancake batter. Adding water—*amending* water, as Mike Oklevitch called it—makes mud thinner and weaker. Overthin mud isn't of much use. That's one of the reasons you can't lay in the rain.

Mike Oklevitch, whose patient, sarcasm-free disposition was almost the opposite of Nate's, and whose response to Scott's approach to

* A manufacturer of natural cement reemerged in 2004, thanks to the market for historic preservation. Speaking of which, palynologists have examined ancient mortars and, from the pollens trapped inside, deduced what plants and foods our ancestors may have eaten.

innovation was opposite that of Rocky, figured out that they didn't have to add water to get mud of the desired workability. They could just mix it. He gained this insight by experimentation, of which Scott approved. Mike Oklevitch tried mixing mud in a food processor. He ran mortar through a meat grinder. He mixed up dozens of batches of mortar, each with a different moisture content. He did this in multiples, and for each batch, he let some sit, some mix slowly, and some mix quickly. Around his experiments, which took place downstairs in the shop, in the domain of Rocky and Erwin and Kerry, he put up signs that said PLEASE DO NOT DISTURB BRICKS, or he surrounded them with yellow police tape. On an old pallet, science was happening, and it was not to be messed with. In each of these mixes, he tracked shear rates, using Kerry's viscometer. He did this to see how workable the muds remained over time.

Running a simple rheology experiment, Mike Oklevitch determined how to take strong, thick, sticky mud and keep it strong and thick and sticky. At first Scott wasn't convinced. Before they had their nozzle configuration dialed, mud did not always adhere perfectly to bricks. Scott would say, *Look at that mud peeling off!* Mike Oklevitch would then take the same mud, slap it on a brick with a trowel, and flip the brick over—whereupon much more mud would peel off. His point: The Stäubli-pump-nozzle combination was already superior to the hand-trowel combination at the command of a flesh-and-blood human. To this day, almost everyone thinks CR's mud is too thick.

In hot weather, mud thickens up fast. It's good for two hours, okay for four hours if tempered, and kicked once eight hours have elapsed. Tempering mortar conditions it. Most guys call tempering shaking, but Mike Oklevitch calls it babysitting. When he works with real mortar, he babysits it to keep it alive. He has learned to keep it workable for six hours, an achievement not incomparable to getting an oil lamp to burn for eight days. It's a subtle art. Mix too slow and mud hardens. Mix too fast and you entrap air, making it fluffy and weak. Add too much water and it effervesces, leaving nasty white streaks that resemble the shit of a pterodactyl. Mike Oklevitch, by sticking his index finger in mud, learned to discern the stuff's rheology. He felt how it flowed.

By August 2013, Mortar Mike had decided to stay full-time at Construction Robotics, and PMD had assembled SAM's frame. Tim Lochner, who once blew off his fingers in a fireworks accident, did the assembly himself. Rocky had said three women couldn't make a baby in three months, but Tim Lochner nearly did it in three days. SAM—one arm, one electrical box, one mixer, one pump—was one of the simplest machines PMD had ever built, even if it was enormous.

As summer turned to fall, stress ran high, because SAM remained framed up but unfinished. The engineers brought it to the shop and put it up on their Hydro-Mobile, such that it was addressing a practice wall. (Veneer brick walls must be braced against something and fastened regularly to it, with small metal brackets called brick ties, otherwise they lean and fall. Scott knew he'd eventually need human masons positioned beside SAM not just to strike the bricks' joints but also to install wall ties every few courses.) The practice wall was made of Plexiglas, so the engineers could see through it and examine SAM's brick placements. Since SAM had not yet placed any bricks, the engineers were left with a perfect view of SAM as they wrenched on it.

Facing the pressure of a rapidly approaching job with a not yet ready product, Scott tried to gain perspective. In his notes from September, he wrote about "must haves/nice-to-haves" and also wrote, "Know when good enough is enough." Other handwritten notes included the reminders "Don't overpromise" and "Don't try to nail a deal on-site." On the one hand, he knew he barely had an MVP, but on the other, he seemed to have forgotten that fact. He figured a machine that paid for itself in two years would be very enticing. He'd hoped to convince the world of SAM's worth by November, but he reassigned his hopes for May.

Easing off the gas was not Scott's style, so the perspective floated away, and stress remained. The stress of trying to keep perspective probably added to Scott's stress. In any case, the stress was so bad that Scott came down with shingles. The virus manifested on his face, leaving him unable to shave—hence the beard. The symptoms were so painful (shingles induces nerve pain) that for three weeks Scott was on Percocet, which left him unfit to drive. In minimally viable health, he stayed home and barely worked half-time, which made the busybody yet more stressed. He looked half dead.

His engineers continued working as if their insistent boss were just around the corner. Late on the night of October 10, after eighteen hours at work, Glenn and Erwin got SAM to place its first brick. It was one a.m. (technically October 11), the latest Glenn had ever stayed at work. He was recording a video of that first (dry) placement when his wife called:

"Hey, where are you?"

"I told you, I'm working late."

Then, because Android phones had a bug, the call shut off Glenn's video camera, so twenty-five seconds of the video were lost. But SAM's motion was so slow that Glenn still caught the second half of the placement. That night, SAM went on to place eight more bricks. The machine put four bricks on top of five, its first mortarless wall.

The next morning, Glenn slept in, and by the time he got to work, SAM was out back near the highway, under a tarp, set up to build six feet of wall against a piece of plywood. It was the first time SAM had been outside, and the first time the machine was laying bricks with mortar. Scott, still in a viral stupor but determined not to miss the occasion, roused himself, limped to the jobsite with pizza and beer, and gave his engineers high fives.

For the first time, SAM did what it was supposed to. All of a sudden, Glenn was running the machine without holding a button or adjusting a dial. Scott noticed Glenn just standing there and was immediately dismayed—as if automation were abhorrent. "Aren't you supposed to be running the system?" Scott asked. Feeling free and untethered, Glenn said, "Not anymore, baby!"

The glory was but brief. On SAM's fourth brick, Scott had Glenn stop the machine so that he could adjust the consistency of the mortar.

Scott adjusted the mud a bit too much, which was evident when the first outdoor wall sagged dramatically, leaving big smiles in every course. Once the mortar had hardened, though, the wall passed a water penetration test—so SAM was deemed job-worthy.

The PMD job, delayed four months, was less than two weeks away.

9.

Refinement

In writing the authoritative book on bricklaying, Frank Gilbreth faced a conundrum. Aiming to promote speed, he tried to portray the craft as simply as possible—à la CliffsNotes. However, after gathering a wide range of artistic tips, he found himself unable to distill them and produced an encyclopedia. He covered, among other things, the Eastern method, a multitasker's technique otherwise known as "pick-and-dip," which entailed grabbing a brick with your left hand and a trowel-full of mortar with your right; and the Western method, which entailed using a larger trowel, spreading out enough mortar for a few bricks, and then grabbing the bricks one at a time. He advised masons to wear hand leathers (finger pads), and to mix lime mortar and then wait two weeks before using it. He offered this tip: To confirm that your corner is plumb, sight your lead to a building across the street (this was useful in the east, less so out west). All told, it took Gilbreth 344 pages to cover everything. It should no longer come as a surprise, then, that a handful of engineers who had neither laid bricks of their own nor read Gilbreth did not have the process dialed—and that the hard work of two actual masons from Syracuse was not enough to save them.

Build, measure, learn: That's what Ries advised. Having done the first two, Scott and his team of engineers spent the remaining six weeks of 2013 and the first six weeks of 2014 in the wood-paneled trailer, collecting their "learnings" from the 1,296 bricks of the PMD experiment. These fundamental units of a *Lean Startup* arrived in bulk. There was so much learning, in fact, that it bordered on soul-searching.

To Rocky, the months of discussion amounted to an extended come-to-Jesus moment.

Scott voiced the overarching question: To proceed, in what direction ought they go? The simplicity of the question belied the complexity of the answers. A few facts were clear: The roller-coaster rails—which were supposed to give the machine maximum access to wall, by allowing it to roll around curves—did not work. (Rocky, opposed to them from the start, called them the worst thing he'd ever seen. He had wanted four wheels, as on any car.) The parallel lasers were excruciating to align. If Mortar Mike was to be unshackled from the position of permanent mud attendant, the machine needed to have its own mortar mixer. It also needed an onboard generator, because managing the electrical cable running from the scaffold to the generator on the ground was annoying. Yet the whole system had to be smaller and lighter. Setting up and calibrating had to be simplified. The machine had to lay bricks more squarely. And, of course, it needed to be faster. Generally, SAM was two orders of magnitude shy of its production goal.

From there, a major subquestion emerged: Should SAM lay brick and block, or just one or the other? Which one? The surveyed contractors had said, clearly and almost unanimously, that while a bricklaying machine didn't especially intrigue them, a block-laying machine certainly did. The industry seemed to *crave* a block-laying machine. But Scott wanted to stay with brick. Bricks, unlike blocks, were admired. The brick industry made sense. Besides, he figured, they'd already come so far.

No matter what you called it—debrief or brainstorm—this period reeked of gloominess. As the need for an overhaul became apparent, Rocky suggested the engineers decamp in different directions and each design the machine of his dreams independently; upon reconvening, they could see what they all came up with. As more snow fell, Construction Robotics was in limbo, and certainly not at World of Concrete. All that made it to the industry trade show in early 2014 was the *rumor* of a bricklaying robot—which was the only aspect of a bricklaying robot that had ever made it to Las Vegas.

Trying to lighten things up, Glenn brought in a barbecue and grilled some meat on a snowbank beside the trailer's front steps. All the while, SAM sat outside like an abandoned car in the parking lot, not fifty feet away. Nobody wanted to touch the beast.

Though Scott had not yet cobbled together an advisory board, he had, during this quagmire, begun informally consulting a future board member named Don Golini. Don Golini, whom Scott met at High Tech Rochester's entrepreneurial class in 2010, had led a life eerily parallel to his own. Don Golini had also attended U of R (class of '86), where he'd studied engineering and met his wife. Like Scott, he'd worked for a while in Boston, then come to Rochester and, at the same age as Scott, started a business. The two hit it off immediately. Don thought Scott was smart, generous, able to talk to regular folks; Scott admired Don's optimism and his technical and business savvy. Don Golini had also invented an expensive machine and had done well for himself. He'd sent two kids to Dartmouth and one to Middlebury and was about to buy a comfortable house on the estuary of the Piscataqua River, where bald eagles would nest in his backyard.

At a firm in Lexington, Massachusetts, Don Golini had made lenses the size of a person. They were for spy planes and the Hubble telescope. To refract light properly, all had to be polished just right. While spherical lenses could be polished with curved pads (anywhere on a sphere, the curve would make full, even contact), such lenses yielded less than perfect reflections, because light does not actually emanate from a single point. Light travels in parallel waves, requiring not a single focal point but a curved one. The surface conforming to this shape is called an asphere. Aspheres—elongated spheres, really—are hard to form and harder to polish. No rigid sanding pad will do, because the surface is not uniform. A rigid pad won't polish an asphere so much as scratch it. Polishing aspherical lenses took skilled artisans, trained to feel contact pressure in certain motions. "That's the history of aspheres," Don put it later. "They're desirable but difficult to make."

Don had wondered if there was a way to automate the polishing. The U.S. Army bought vast quantities of American-made binoculars, each with aspherical lenses. Automating the polishing would open up a huge market. He had no idea if it was possible, but he wanted to try.

In an antiquated Belarus laboratory, Don found magnetorheological fluid. It was a special slurry the consistency of motor oil that, in the presence of a magnet, became as solid as set mortar. Magnetism linked the fluid's chemical chains. In magnetorheological fluid, Don saw an adjustable (reversible, even) polishing paste. More magnetic field meant more viscosity, and more viscosity meant more cutting rate. A paste with an infinitely changeable cutting rate would far outperform the imperfect artistry of people polishing glass—at least that was Don's theory. Sometimes that theory seemed worthy. Other times it seemed worthless. Plenty of others doubted him. He doubted himself.

It took Don years to develop magnetorheological-polishing technology (he, too, built his own viscometer and hired a MacGyver type as his first employee), and two more years to build a machine that employed it, before he had found his first two customers. They were local. For about the cost of SAM, Don sold three machines that first year. He sold six the next, then twenty. After that he stopped taking investments, stopped giving up company equity. His company grew to seventy-five employees. Then he sold the firm and became an angel investor.

Through a mastery of viscosity, Scott also yearned to sell expensive machines to an industry that had long made do without such things. From Don, Scott sought wisdom. Don, through a thick Boston accent, obliged.

Don thought the most important trait Scott could focus on was self-awareness, because "the genes that contribute to starting a business don't necessarily go with the genes that contribute to humility." Don knew he was assertive, more so than Scott. He also thought Scott should prepare himself for ambiguity and uncertainty, situations not generally well handled by an engineer. Don reminded Scott, by way of encouragement, that he couldn't put too much weight in market research, because Construction Robotics was offering a product that people didn't yet know existed. Of course people wanted a block-laying robot—they also wanted flying cars!—so what kind of answer was that?

Scott showed Don his company's books and explained the crossroads they had come to. Their prototype was such a clunker that everyone wanted to redesign it, but they had little money and no imminent revenue. Don, speaking frankly to the engineers, delineated their choice. He said the guys either needed to move forward with SAM as it was or shut down. The situation, as he saw it, was sink or swim. There was no wiggle room.

Sitting in the trailer, the engineers did not bother to vote. The choice was obvious. SAM, they knew, was not up to any bricklaying job. They agreed to shut down.

Rochester is a city with a long history of innovation and reinvention. When the Erie Canal came through in 1823, linking the region to the east, it transformed Flour City (named for the abundance of flour mills) into an industrial boomtown. George Eastman arrived in 1860, attacked a generation-old technology, and soon made photography so accessible that captured images took on the company name. "You press the button," he said. "We do the rest." In 1926 alone, Kodak used two hundred tons of silver and produced enough film to circle the planet eight times. That same year, Eastman put into words his vision of constant innovation: "A company that contents itself with present accomplishments soon falls behind." That might have been Rochester's motto. The next year, using part of the abandoned Erie Canal bed, the city started a subway system that ran for thirty years. It ran from GM and Kodak, past the University of Rochester, southeast almost to Victor. The subway survived until the arrival of the interstate, which filled up with automobiles made by General Motors. From where SAM sat in the parking lot, you could have thrown a brick and hit the highway. Rochester was where Bausch + Lomb made everything from microscopes to Ray-Bans, where Harris made tactical radios, and where Xerox made possible such mountains of paperwork, every sheet an eponymous creation, that it was almost orogenous.

Nate remained a believer in automation, as he had been for most of his life, largely because he had so much experience with doing things the old-fashioned way.* Before bar codes were invented, he'd stocked supermarket shelves. Before pressure-treated lumber was available, he'd soaked boards in Cuprinol. Before scanning machines started sorting postal mail, he'd stuffed envelopes in pigeonholes. But his first experience with automation had come even earlier, in the summer of '56, when, at nine years old, he stepped into the bowling alley that his father had bought.

The place was called White Barn Bowling and had been established in Brewster just after the war. It had six lanes and, like most bowling centers, relied on pinboys. Pinboys (also known as pinaroos, pinsetters, or pinstickers) were kids who worked in the pits at the ends of the bowling lanes and reset the bowling pins after every roll. At most lanes, they kept games going by placing the pins, which had holes in their bottoms, on spikes that could be pushed up from the floor of the pit. In fancier lanes, they set—or chucked—pins in the pockets of racks that descended from above. These pinboys were also known as woodchuckers, since they were chucking wood. Pinboys at old and new lanes were always on guard: Strikes would send pins flying, and perched on the back edge of the pit, they sought to reset the pins as fast as possible and send the ball back to the bowler. By picking up three pins in one hand and one or two in the other, a pinboy could reset a lane in three pickups. Each pin weighed three pounds, so it was work—but it was better than delivering papers or groceries. It was exciting. It was also dangerous. Bowling balls, which weighed up to sixteen pounds, were "ebonite missiles" and required astute dodging—especially if a pinboy was setting two lanes at once, which was common. Pins from one lane could fly sideways and smack a pinboy in the next-door pit. Pinboys endured smashed fingers, banged shins, even broken ribs. Some wore shin guards to protect their ankles and still had to play dodge. Drunkards were the

*Some irony: As drawn as Nate was to the new and improved (long before computers were mainstream, he urged Hueber-Breuer to invest in them), he remained fond of the old and traditional. Painstakingly, he built wooden boats and much of the furniture in his house by hand. He missed the days when buying cheese led to a conversation with the grocer rather than the impersonal beep of the bar code. Yearly, he went fishing near the headwaters of the Kennebec River, always staying at a Wi-Fi-less, cell-service-less, off-grid lodge that, for most of its refrigeration, uses two-hundred-pound blocks of ice cut from a nearby lake every January and stored in cedar shavings.

most dangerous, as were jerks who launched rockets or threw two balls. To get back at such wise guys, a pinboy would return the ball with backspin, so that it stopped before reaching the bowler, forcing him to duckwalk out the lane and retrieve it. Some bowling establishments provided insurance for their pinboys, since some of them would inevitably be carried off to hospitals.

A pinboy who avoided injury could make decent money, though. In addition to his wage of ten cents per game, he earned tips, which often came in the form of a quarter or half-dollar rolled down the lane. A lucky pinboy might find, rolled up in the finger hole of a slow-rolled ball, a dollar bill. A well-tipped pinboy could generally be counted on to give a wobbling pin a nudge in the bowler's favor. If he set two lanes five times over the course of an evening, he could take home a dollar in wages and a couple more in tips. If it was a league night, and he worked his tail off, he might emerge from the pit drenched in sweat but with eight dollars in his pocket. And if he worked all day (some, over holidays, worked thirty-six hours straight), he might set 125 lanes. A pinboy of such stamina might get invited to accompany a bowler to a tournament. Often owners let pinboys stay until sunrise and play a few games of their own.

In a hundred-point game, a pinboy's work amounted to lifting four tons. Even kids run out of energy. And they had curfews, which limited the hours a bowling alley might operate, especially during the school year. So when Philip Podkaminer and his brother Leon opened Brewster Bowling (now with ten lanes), they immediately installed automatic pin-setting machines.

The pin-setting machines were made by American Machine and Foundry, the same company that would soon make one of the world's first industrial robots. At least half a dozen tinkerers had schemed up automatic pin setters, but one, made by Fred Schmidt, held appeal—even if it was made of car parts and lamp shades and flowerpots and bicycle chains. Brunswick hadn't been interested, but AMF was. In fact, the company was on a mission. Since 1941, when Morehead Patterson inherited AMF from his father, he'd been looking for large unmechanized markets to invade—and bowling, reliant on so many pinboys, nearly screamed at him. Schmidt's Model A was eight feet tall and weighed two tons and made a lot of noise, but it used suction cups rather than human hands and even had a "pindicator." It was the future. Still, AMF had some questions: Was it fast enough? Reliable? Affordable? Repairable? Would it change the game? Might it be approved by the American Bowling Congress?

To find out, AMF tested the Model A pin spotter at the 1946 ABC National Championships, in Buffalo. In an armory three blocks from Lake Erie, AMF demoed two of the machines for two months. AMF kept a publicist and a salesman on-site—and when one of the machines went into "self-destruct mode," the crew hurriedly switched to the other machine. The company gave out buttons. All the national networks came out—Paramount, MGM, Universal, Fox. Since there was no such thing as a zoom lens yet, a Paramount man removed his shoes and ran down the lane, pushing his camera on a dolly right behind a bowling ball. Columbia made a ten-minute short featuring a few bowling champions, narrated by Bill Stern. It was called *Ten Pin Magic.*

Georgia Veatch, the editor of *The Woman Bowler*, called AMF's automatic pin-setting machine "Johnny Amfaps," and wrote:

> I never thought I would see
> A robot as marvelous as thee—
> Of thee Rube Goldberg might have dreamed,
> Or on whom Franklin would have beamed . . .

By the end of the demo, AMF had orders for ten thousand machines, and the company knew it was onto something big. But the Model A was imperfect and too expensive—so AMF scrapped the design and started fresh. Thus 1946 turned into 1947 (the year of Nate Podkaminer's birth) and 1947 turned into 1948. The redesign continued.

It was during the summer of 1952, from a three-hundred-thousand-square-foot plant in Cheektowaga—less than twenty miles from Akron—that AMF began producing Model Bs, spitting them out like pretzels. Crucially, the company had decided to rent (rather than sell) pin-spotting machines—and, with these improved pin-setting machines, it changed the sport of bowling.

As suburbs spread, bowling alleys nearly sprouted from the earth and became the country clubs of the middle class. Bowling—once thought of as debauched for its connection with pool halls, gambling, and crime—spun off and gained a reputation as wholesome recreation, suitable for men, women, and children. Bowling was cheaper than the movies, safer than roller skating. Bowling came up out of the basement. Before long, it was big. In the mid-1950s, there were eighteen million American bowlers, more than the participant count of any other competitive sport. Two million Americans belonged to leagues. Analysts figured Americans bowled two hours a week. Charles Schwab, new to Wall Street, was compelled by the industry, which grew to something on the order of a billion dollars. AMF stock doubled between 1957 and 1958. By 1960, there were more than twelve thousand alleys in the country, each with an average of nine lanes. Because the overwhelming majority of those establishments had pin-setting machines, which delivered consistent cycle times 24/7, pinboys were out of work, but wheel makers in Youngstown were employed, as were motor makers in Fort Wayne and electrical engineers in Chicago. That was the angle AMF took, anyway; by then, their motto was "Electric pinboys never go home." Freed from the constraints of manual labor, Philip and Leon Podkaminer kept Brewster Bowling open from noon to midnight, seven days a week.

Nearly sixty years later, Nate remained a firm believer in automation, and he came to the rescue with some investors. He had rounded up some funds a year before, when the cost of building SAM ballooned (the Stäubli arm alone cost as much as two years at Stanford), and this time, from friends and family, he gathered even more; he was not about to give up. Because neither he nor his son-in-law intended to give the company to outsiders, they were able to swallow this move. Scott, in particular, was determined not to give away the company to the worst kind of outsiders, the venture capitalists in California. He despised the VC hype, hated that so many "entrepreneurs" rubbed shoulders all day long with VCs and never actually made anything. He held a grudge and admitted that it drove him. He aspired to prove SAM's worth without the Silicon Valley VCs. Venture capital money might have been green, but to Scott, it was poisonous. He wanted to maintain his modesty, remain the underdog, and make things the way Americans once did—the way his grandfather had. He wanted to thrive without the hype—without Burning Man—and end up more successful than people realized. As such, with friendly funding, the redesign commenced, and the Alpha machine was dropped as the Beta machine was developed.

Don Golini, who had found insights in *The Innovator's Dilemma* and *Crossing the Chasm*, knew that Scott had to devote as much effort toward sales as to engineering. Per MVP dictate, Scott needed to push SAM before it was ready, at peril to the company. To Don, it was an especially urgent matter; even though the construction industry was vastly bigger than the optics industry, he didn't see the demand that he'd seen in optics. He reminded Scott that the role of a CEO was threefold, in this order:

1) Sales
2) Recruiting
3) Knowing when the cake is ready

The problem was, of course, that the metaphorical cake had not been put in the oven. They didn't even have a recipe.

Nate was of the same mind—Construction Robotics had to start selling *now*. So he hired a civil engineer from Syracuse named Chris Raddell to lead CR's sales effort. Nate, on the board of the YMCA in Syracuse, had met him through the Y's executive director. They were neighbors. It turned out Nate and Chris already knew each other, from a backyard high school graduation party.

Chris Raddell had grown up in Cleveland, five houses away from Lake Erie, and, like Scott, spent most of his childhood in or on the water. In the summer, after storms rolled through, he used to find old telephone poles and paddle them a thousand feet out to freshly formed sandbanks. He got a rash from the pole every time, but it was worth it. In the winter, he would carry a toboggan to the crest of the fifteen-foot ice jam on the edge of the lake, then ride the frozen moraine inland, holding on for dear life because the ice was chunky and sharp. Once, he hopped onto a block of ice and headed north, into the lake, using a pole for propulsion—and made it twenty feet out before a Coast Guard helicopter flew over and yelled at him: GET BACK TO SHORE. He might not have been a long-distance swimmer, but he was an engineer in the making, and he never lost his childish excitability.

In high school, Chris had helped out at his father's butcher shop. His dad inherited the shop from his father, who was born Ludwig Radelj in Slovenia and tried mining lead in Colorado before landing in Cleveland, which had appealed to him because it had more Slovenians than Slovenia. Neighborhoods in Cleveland followed a pattern: Every six blocks, there was a barbershop, a salon, and a butcher shop. At his dad's butcher shop, Chris started out doing the grimiest task: cleaning casings. Casings are pig intestines. They showed up in barrels, packed in salt, by the hundreds of feet. Every Saturday morning, starting at four-thirty, Chris spent an hour or two running water through twenty-foot sections, thinking the whole time: *The second-to-last thing that touched this was shit.* He wore an apron made of plastic over another apron made of cloth, and still came out covered in salt splatter. It was the worst job at the butcher shop, and to Chris, the work seemed like a penance. He went through a lot of bleach. His father, meanwhile, wore a tie to the shop and looked like a professional. Chris repeatedly asked his father, *Can't we buy prewashed casings?* His father, who spoke Slovenian but didn't teach Chris the language because he wanted his son to fit in, insisted that they make sausage the traditional way, as so many generations before had. It was, he knew, a good business, and it allowed him to raise eight kids.

From casings, Chris worked his way up to cleaning the floors. From there he rose to general cleanup. Thence to sausage making. He learned to mix spices in with pork, stuff the casings, and rack thousands of sausages for the smokehouse. If you draped two links left, then two links right, you could get fifty links per wire and four or five hundred pounds on one rack. Because smoke wouldn't stick to water, Chris's first session in the smokehouse merely dried out the meat. After that, using handfuls of oak and cherry sawdust, he learned to produce smoke for up to eight hours. Firemen used to come investigate regularly.

In the weeks preceding Christmas and Easter, the Raddells squeezed in two smokehouse shifts per day by starting before dawn and working until three in the morning. In this way, they could produce ten thousand pounds of sausage in a week. The work nearly killed them, but the sales made it worthwhile.

The next rung up from making meats was cutting them. The top of the ladder was selling, and talking to old ladies. Old ladies would point their bony fingers at the particular sausages they wanted, saying, *I want that one right there*, and Chris would oblige. Through this, he learned to talk to people and sell them things.

Then he learned to make them, by training as a civil engineer. After college, Chris got a job at Chicago Bridge & Iron, which made pressure vessels for the nuclear power industry. The nuclear power industry, reliant on so much new technology and so many new procedures, and the target of so many powers-that-be, promptly crashed.

The next best thing for a civil engineer was life at Parsons. Parsons doesn't just build buildings. The company builds buildings in challenging places, and builds them securely, even secretly. Parsons built the United States Embassy in Russia after the first go was so full of bugs that the U.S. government refused to use it. Parsons brought water to Las Vegas—no small amount of plumbing—and floated equipment to Prudhoe Bay and assembled it long before the Alaska Pipeline was built. Parsons was the company the government went to for a new embassy in Iraq, and it's the company extending the D.C. Metro that Bechtel built.

Chris Raddell spent twenty-seven years at Parsons and, with his security clearance, traveled to sites for six weeks at a time, contributing to proposals a foot thick. With Chris Raddell's help, Parsons cleaned up the World War I chemical weapons (including Lewisite, aka the "dew of death") buried in D.C.'s Spring Valley neighborhood, aware the whole time that the downwind zone included the White House. The government granted the company a special indemnification in advance of the work.* Chris became one of Parsons's one hundred

*Apparently, it did not bother the government that Parsons, on the government's dime, had built the factories in which those same chemical containers were filled sixty years earlier.

vice presidents: the VP of business development for infrastructure and technology. While he occupied the role, the company enjoyed 10 percent growth for a decade. He rode in helicopters to Hawaiian Islands and had meetings with higher-ups from the air force. He visited all fifty states, racking up a million and a half air miles.

Then Parsons tried to relocate Chris to California, far from his church and three sons. A parish trustee and the head of his church's finance committee, he said no thanks and found work with SRC. SRC had made a quarter-million-dollar device that blocked cell phones from triggering IEDs, and needed help getting into this billion-dollar business. It had also developed a radar so fast and powerful that the fellow operating it could detect a fired mortar shell, radio a Black Hawk pilot, and have him bomb the launch site before any more shells could be loaded and launched. Business in this arena was good—until war funds dried up, at which point Chris Raddell became an underemployed independent consultant.

Then he met Nate. To Nate, Chris Raddell was perfect: He understood family business (where "a job is a job until you're done") and was familiar with all varieties of construction. He knew sales, and engineering, and even new products. He didn't need much sleep (he had gotten five hours a night for the length of his career) and didn't want fame and acclaim. He was happy doing deals behind the scenes, free of the layer of micromanagerial fat found at places like Parsons, and happier still to pitch masons and masonry-firm owners rather than government bureaucrats, which required so much politicking. He liked the perspective he'd have at a place like Construction Robotics: not so low that he'd get lost in circuit design, but not so high that he'd have to steer customers looking for details to more knowledgeable engineers.

He had put together and run nine-figure contracts—contracts for oil companies—which meant that businessmen trusted him. He was, in words that came from him but may as well have emanated from Nate, a doer. He was nerdy for sure, but innocent and forthright, and kind and open, as gray-haired and ministerial as Mortar Mike, as fast-talking as Scott, and liable to say something like "I'm the son of a butcher. Guys call me an SOB."

Chris Raddell was liable to say one more thing, about meat: "Most sausages are average or garbage." He knew, because at the same butcher shop where he'd grown up working, sausages still emerged at the hands of his brother. There was a slight difference, though. A dozen years before, his brother had gotten an automated smokehouse. The new smokehouse was twice as fast and took far less effort than the old way, allowing him to shuffle through three racks daily, and never bother any firemen.

10.

The Team

Concurrently, marketing and building began. On the marketing side, Zak began building a company website, but the task depleted him and nearly destroyed his spirit. Language did not come naturally to Zak, who once described himself as having a learning disability in reading and writing. Spreadsheets came to him more easily than sentences. Throughout middle school and high school and into college, he'd sought tutoring in English while his classmates studied foreign languages. He knew the vision shared by his father and brother-in-law, even knew construction, but putting all of it into words thwarted him. Even writing his own bio stymied him. For years, occasions that called for press releases would summon dread in him.

Meanwhile, Chris Raddell and Scott headed south to push their product in a different way. In meetings with firms in Pennsylvania, Ohio, D.C., and Virginia, they told executives they'd gotten a robot to lay bricks. It didn't hurt that the *Democrat and Chronicle* said SAM could lay three thousand bricks per day, or that *Engineering News-Record* had put SAM in its pages. Never mind that the Alpha machine, laying bricks on the PMD building, hadn't been so much working for a superintendent as showboating for a kind benefactor, or that the machine was currently being disfigured and overhauled. The future was now real enough to read about, and Scott wanted to see who found it enticing. If he made his bricklaying robot easy enough for a dog to set up and use, he asked, did anybody want it?

One of the executives they met with was the head of one of the biggest of the big boys around: Clark Construction. One of Nate's colleagues at Hueber-Breuer had worked at Clark and made introductions. Clark didn't perform masonry—the firm always subcontracted it out—but Clark's CEO loved the idea of SAM so much that he could imagine eventually buying dozens of machines (advertising, all the while, that Clark was the type of progressive firm to invest in the most sophisticated technology) and then encouraging their various subcontractors to use them. Until that was possible, Clark offered to support Construction Robotics in every way it could, showing Scott and Chris Raddell upcoming projects (to see if SAM might be incorporated), introducing them to masonry contractors in the D.C. metro area, and offering the assistance of their own research division. A construction firm with a research division was a rare thing, sure to provide a boost, and this early, overwhelmingly positive meeting injected great optimism in Scott's blood.

But another executive at a different construction firm, catching more than a drift of doom, stormed out of his meeting with Scott and Chris Raddell, saying, *This is bullshit!*

In Victor, good news arrived. With the PMD building complete, their former quarters were now available—so after fifteen months in a trailer, the engineers finally got a regular office. The space, however, was regular at best. The wide gray two-story building sat so close to the interstate that, from inside, you could hear the engine brakes of westbound trucks slowing down for the Victor exit. Upstairs on the west side of the building, the sloping ceiling was so low you could reach up and touch it where it leaked. Downstairs, on the east side of the building where Kerry and Erwin were stationed, different leaks dripped into buckets. A visitor would have had a hard time finding either of these spots, because it wasn't obvious that Construction Robotics occupied them. There was no sign out in the parking lot, no reception area. Past a yellow birch and a flagless flagpole on the north side of the building, a glass doorway led to a tiled room with a copy machine and a safety-glasses dispenser and a display case full of metal molds for making Styrofoam trays. There was no brick or robot in sight. To the left was a half-bathroom/half-kitchen; straight ahead was a set of double doors; and to the right were a few cubicles and offices. It smelled like cutting oil. If you went through the double doors, you encountered massive milling machines run by the company that made the Styrofoam molds. The molds were used by McDonald's and Taco Bell, and to form five-part lunch trays, and for the trays on which supermarket chicken breasts were packaged. Erwin and Kerry and CR's "shop" was seventy-five steps down the polished concrete, past the milling machines and forklifts and pallet racks, but it's unlikely a visitor would have ventured so far that way.

To find CR's offices, a visitor had to proceed to the right, around some cubicles that also belonged to the mold-making company, and find a steep mortar-spotted staircase in the corner. The staircase felt funny because it wasn't to code; the treads were too short. A visitor would have found it convenient to turn his feet sideways on the way up, and proceed with caution on the way down. But the mortar spots were the first hint. The newspaper clips on the wall were the second. The bricks on the windowsill atop the stairs were the giveaway. Yet a visitor arriving upstairs would still feel unsettled. He'd find himself between two conference rooms: one with a perfect view of highway traffic but for the yellowing blinds and window frames bespeckled with mouse turds, and the other featuring a huge dry-erase board and—thanks to Scott's incessant planning—no fewer than three calendars. Straight ahead, he'd pass a watercooler with no water, three offices on the left, and an open area on the right. One of the engineers here—Tim, Glenn, Mike Oklevitch, or Rocky—probably would have greeted him. From there, the visitor might have heard beeping forklifts emanating from the floor. If he took a seat in Zak's corner office, he might have heard groans, farts, flushes, or even turds splashing from the bathroom below. Only Scott's office, next door, offered relative silence.

To begin the overhaul, Scott hired a few more guys. Aside from the cases of Zak and Chris Raddell, all hiring fell on Scott. In engineers, he was looking for eagerness, fearlessness, a desire to work hands-on, and preferably some attitude and spunkiness. He had no interest in yes-men. He wanted a company where everybody was willing to challenge everybody, full of personalities that were tolerant of dynamic situations. He had no test to measure this last trait, but he thought his gut could discern it.

Serendipitously, just as Scott began the hunt, he got word from a few people that he ought to hire a power electronics expert and Harley-Davidson rebuilder named Paul Spronz. Paul had just been laid off, and at Construction Robotics, his mission would be to figure out how to cram all of the Alpha's computers and wiring into a much smaller cabinet.

Soon after, Scott hired another motorcycle aficionado named John Nolan. Nolan wasn't a Harley guy, though. He rode a mid-'90s Yamaha and a bright red Suzuki SV650. Usually, though, he drove his wife's Volvo. He did not, as you might suspect of an engineer, occupy a narrow band of the world. He was an explorer, an avid museum visitor, and a volunteer firefighter—the best traveled of the engineers at CR. Having been to Thailand and China, he once described Thai potstickers to Mike Oklevitch as "Asian pierogies." He had a beef with American trains, not just because he had gathered enough information from friends who worked at General Railway Signals to conclude that the American system of signaling was nuts, but because he had ridden European rails. On his phone, he kept track of the weather in London and Paris.

John was the kind of guy who brought his own tools to work. He had a sense of humor, but it took some coaxing to get there. He found comics, put them on Glenn's desk, and around the office they went. They were usually something snarky, like *Dilbert*. Sometimes, to put a situation in context, he mentioned an old line from his firefighting days: "Chief's here. Everything's fucked." On his helmet, which was white, his surname was printed in all caps via label-maker. He walked stiffly, a little bit like a nerd, a little bit like pride supported his back. But if he carried an air of superiority, it was warranted. Scott recognized that John had a "strong personality," but ultimately respected him and called him a great engineer. He said this in part because he knew John harbored no fear of the unknown. "Some people react with 'Oh my God, there's all this stuff I don't have any clue about,'" Scott said, "and John can have an intelligent conversation about it."

For the last generation, nearly everyone in John's family had worked at Kodak. At the largest industrial site in America, they all worked in different buildings: 328, 64. The last to leave was his uncle, at the very end. John, who'd gone to Purdue, had steered clear of that tradition. Ever the beer connoisseur, he'd interned at Budweiser. Then he'd gotten a job at GM's fuel-cell lab—where he soon sat near Scott, Tim, and Glenn. In ten years, he'd tested a lot of batteries, examined the frame of the Chevy Volt, and calibrated flowmeters in an old missile silo.

When GM offered John the chance to relocate to Detroit, he was willing, but his wife was not. So he found a job at Calvary, another automation company much like the place where Tim and Tom worked before they started PMD.* At Calvary, John spent most of his time working on three huge machines: a four-way-seven-way connector for the big three auto manufacturers; a machine that, with eight hundred thermocouples, made Gorilla Glass for Corning; and, for BorgWarner, the giant producer of turbochargers for Indy cars, a machine that made adjustable cam phasers for modern combustion engines.† In Juárez and Ningbo, John wrenched on these contraptions. "It's all the same nonsense," he said. "Robots, conveyers, welders—for the most part, you have no idea what the machine does." John stayed at the company a year, until he couldn't handle the iron fist of the owner any longer. To him, the owner seemed an IKE: an "I Know Everything." John was amazed Calvary hadn't gone under, and when he left, he certainly wasn't bothered that he started a mini-exodus to PMD, one of its competitors.

GM had two hundred thousand employees. Calvary had two hundred. Scott's start-up had a dozen, and to John, who was sick of making crap for other people, this sounded good. He knew Scott and was happy to do what needed doing, no matter how extensive.

*For someone eager to make a $10 million machine, Rochester offered still more choices: Alliance, Liberty, and Maris.

†BorgWarner once made three-speed automatic transmissions for Studebakers, which connected John's predecessors to Nate Podkaminer's father, who, before Brewster Bowling, had owned a Studebaker dealership.

He started without a title. From day one, CR demanded a lot of John, but he never panicked. Amid a steady barrage of engineering requests, he blocked off his weekends, scheduled vacations, headed to the Thousand Islands, even turned off his phone. Engineering problems, as he saw it, didn't really change and would go on with or without him. On the small wall in front of his desk, he pinned up a periodic table of sports cars, divided into noble supercars and poseur-oids.

His confidence emanated in comments like:

Glenn: "How do you know it's a Stäubli problem?"

John: "It has to be a Stäubli problem."

Starting from scratch, the engineers embarked on an iteration stage that followed in equal proportion to the learning stage. Abandoning the roller-coaster rails, the engineers relinquished their aspiration to build a machine that could lay bricks around corners, and instead conjured a simpler machine whose proficiency would be straight flat walls. Toward this end, Tim began designing a sturdy, flat, two-foot-wide steel platform that, in sections, could bolt onto the deck of any Hydro-Mobile. Like a bike lane on the side of a street or a boardwalk on the edge of a beach, it would leave the main area clear for all kinds of haphazard activity, while reserving a narrow causeway for back-and-forth motion. The other guys spent two weeks looking at smaller robotic arms. Glenn produced an analysis of the options, weighing speed, payload, and ease of controlling the robots. As for mortar, after contemplating one arm for mortar and one for bricks, using a hose to squirt mud on the wall, and applying mortar in-line (as each brick was conveyed to the gripper), they began closing in on the hopper they wanted and what mixer they wanted to put in it. As for the pump, they still weren't sure if they wanted a single or dual augur, so Mike Oklevitch began testing.

As for the metrology system, Rocky had an idea. Two lasers had been a pain in the ass, especially because making tiny adjustments (by rotating a pair of set screws) had a huge effect sixty feet away. Contrary to the opinion of RPI, Rocky realized they could get everything they needed from just one laser. If they kept the time-of-flight laser but repositioned it at their desired course height and wall offset (providing

X), they could also use a camera to look at where the dot was reflecting and thereby figure out Y and Z, their last coordinates. X, Y, and Z marked a spot, right? A single laser could perform two tasks at once: provide distance and position. Dollar-wise, getting rid of the fiber-optic bundle and sensor and replacing it with a camera would be a wash, but in terms of ease of use, it would be a huge gain. Quickly, this new almighty laser was deemed a Smartline, and it looked better than a piece of string.

The guys also figured out that instead of setting the height of every course manually, with one man standing on the left with a square, and another at the laser box on the right adjusting the angle of attack, they could set up the system once and make all subsequent course adjustments mechanical. If their system used two poles, each with adjustable tabs, the laser box could climb up one and fire its laser across to reflectors on the other, leaving the men with nothing to do but sit and watch. They'd just have to mount the poles, set the tabs, and then voilà!

Rocky found all this scheming a joy. For the fourth or fifth time, he told Scott that he'd work at Construction Robotics for free. "You better be careful what you wish for, Rocky," Scott said, because his scrappy start-up still had no revenue.

Glenn suffered more doubts. Regularly, he wondered if any of their calculating and tinkering would work. He was a linear thinker, not used to leaping as far ahead as Scott. But he had faith in his boss, recognizing that he'd never been exposed to a visionary before. You don't get visionaries at well-established companies. Years later, he identified similarities between Scott and Steve Jobs. Scott made radical requests because he knew where things could go.

John quickly figured out the family ties at Construction Robotics and began to detect unusual harshness toward Zak from Scott and Nate. To John, it seemed not just tough love or anti-kleptocratic resistance but more. That wasn't to say John and Zak were exactly chummy; in fact, John found Zak irritable and unfriendly and, having endured a stream of grating interactions, mentioned the dynamic to Scott. Scott told John not to take it personally, but from nearly day one, John and

Zak engaged in regular spats. The cultural differences between the two men made it inevitable.

John watched art movies. Zak played Call of Duty. John read critics. Zak read the *Drudge Report*. John had marathons under his belt and still liked to get up early and go on long runs. Zak thought running was dumb because it was like soccer with no ball; since damaging his knees, he had stopped playing lacrosse, taken up golf, put on fifty pounds, and did not operate on a construction schedule for fun or for work. John was a planner. Zak regularly woke up less than an hour before flights departed and maintained that he always arrived at the airport with "room to spare." Zak texted while driving and said he was better at it than most people because he had a lot of experience doing it. Straddling a motorcycle, John knew to look out for drivers like Zak.

Zak overworked himself to exhaustion, often staying at the office past midnight and going to bed without dinner. John knew the work was infinite, so he meted it out and made it home nightly for dinner with his wife and kids. Except in matters of automobile cleanliness, Zak was not fastidious. John certainly was. He and his fellow engineers needled the only non-engineer in the office, calling him the head of the HR department.

A paragon of health Zak was not. "All I eat is eggs, ice cream, and cold cuts," he once declared. He was forgetting pizza, Monster Energy, and canned beer. He tended to put so much sugar in his coffee that a) it cooled off so fast he could chug it and b) he was left with a thick sugar sludge at the bottom that he spooned out. He consumed Rochester's specialty, the "garbage plate": chili, piled atop a hamburger, piled atop french fries, piled atop macaroni, piled atop beans. Was he on the Atkins diet? No, he said, "I'm on the I'm Fat diet." Barely thirty-one, Zak was, in his own words, as flexible as a ninety-year-old. He was seventy-five pounds overweight, the owner of a belly his niece once mistook for that of a pregnant woman, and saddled with knees that did little more than inform him of incoming rain.

John's food preferences were healthier and less rigid. All ethnic foods held more appeal for him than hoagies, all microbrews held more appeal than canned swill, and he preferred not to revisit the same restaurant three times. The garbage plate did not enter his life. Where Zak slammed a shot of Oban from a red plastic beer cup as if it were crappy tequila, John sipped his liqueurs.

Zak, who identified as a Republican, listened to classic rock and didn't recycle. John, who identified as a liberal Democrat, listened to podcasts and did. John read books that got reviewed in the *New York Times* and kept them on bookshelves. Zak had no such bookshelf of erudition, but he had a copy of *The Millionaire Next Door.* Zak, though, possessed more humility than John. "I'm smart," he said, "but not that smart."

At the office, all that separated the men was a door and a low glass wall. About the only things they had in common were smarty-pants in-laws. John's sister-in-law studied Ebola at Harvard. Zak's brother-in-law was trying to build a bricklaying robot.

The engineers brought their old prototype, the Alpha machine, into the shop downstairs and started hacking away at it. Meanwhile, Scott tasked Glenn with redesigning SAM's map tool. Glenn, knowing how robots work and now having some sense of how walls were built, needed to make it easy for a mason or foreman to design and adjust walls on an iPad—to adjust bonds, set window placements, check course alignment a few stories up. In essence, he needed to build a modified Tetris. This being no small task, Scott hired another engineer to program the new, smaller, bright red Stäubli arm they settled on.

Back when Scott was reading start-up books, a student named Chris Johnson had considered trying out for RIT's baseball team but figured it made more sense to focus on engineering. He didn't want to get hung up during his freshman year and fall behind, not for the glory of second base, not when his parents were paying for college, and especially not when he intended to pay them back. His dad had been a chemical engineer at Kodak, and when he was laid off, he advised his son to pursue a different field. So Chris Johnson studied robotics. He learned to compute inverse kinematics for five axes, and learned that the human body, technically, had something like one hundred and eighteen axes. He wrote his thesis on RIT's TigerBug hexapod, which bore a freakish resemblance to a large tarantula, except each leg had three servo-driven motors. Another hexapod, the PhantomX, particularly impressed him. Actually, what got him was the code it was running on.

Now Chris played in a Thursday-evening baseball league; he played basketball, too. Slowly, he was ticking off all the four-thousand-footers in the Adirondacks. Primarily, though, he spent his free time lifting weights—and there was no way to overlook this. Chris's biceps were bigger than Glenn's quads, bigger than Zak's head. Because his heavy brow and prominent chin gave him the kind of head that could be in a Russian mural, and because, when lost in thought, he twitched his thumbs as if bursting with energy, his overall composure suggested nothing so much as a superhero. Everywhere he went, he brought a green sports water bottle, as if he had been working out just around the corner and had to hydrate and recover while coding. He drank protein shakes, and the headphones he wore were so big and green it looked like they'd been drinking protein shakes to bulk up, too.

More so than any other employee at Construction Robotics, Chris Johnson was young and free.* He didn't wear gloves when working with mortar the way Mike Oklevitch did, and he had holes in his right boot. He sometimes sat on his desk. While working, he listened to music or podcasts and occasionally hummed Taylor Swift songs or the theme song from *Jeopardy!* Where Glenn and John had photos of their kids on their walls, Chris Johnson had an espresso machine on his left, a coffeemaker on his right, and above his laptop, a cartoon taped to the wall. The cartoon featured an engineering joke—no, worse: *a debugging joke.*

"What should a coder do in case of fire?"

"First save your code. Then leave the building."

On that same wall, he'd also posted the six stages of debugging:

* Another Chris Johnson, who *has the same birthday* as ours, is a felon—and to the TSA personnel who sometimes confuse the two, our Chris Johnson has said, "I'm not in prison, and I'm not black!" To which the TSA personnel have said, "We don't look at skin color, because that's discrimination." This only further infuriates our Chris Johnson, whose job it is to process logic.

1. That can't happen
2. That doesn't happen on my machine
3. That shouldn't happen
4. Why does that happen?
5. Oh, I see
6. How did that ever work?

Chris Johnson loved robots, and he loved Construction Robotics, mainly because he got to mess with SAM's big red Stäubli arm all day. The quantity of work did not faze him. Where his roommate's corporate job yielded a normal schedule and a reasonable workload and better pay, Chris's start-up path entailed working through back-to-back weekends to solve god-awful equations with no help from Stäubli. Even when he left work, work did not leave him. He had nightmares filled with unsolvable equations. Once, when he was supposed to pick up his roommate at the airport, he told Glenn, "It doesn't matter. Let's just finish this." He stayed put to resolve a bug, which he found a thrill. To Chris Johnson, the thrill was so great that, improbably, he called bugs "the best things ever"—because the result was understanding. "Bugs are awesome," he said.

As Chris Johnson found bugs in the robot code, he filed them away in categories in his head. After a few months, he had enough evidence to conclude that SAM—or at least the Stäubli arm—was a piece of shit. Like Scott, he dreamed of a fast, elegant machine.

Once, while writing the code for the arm's motion, which he had broken up into thirty smaller functions, one particular snippet that hacked into Stäubli's servo-loop stymied Chris Johnson's progress for so many weeks—producing not a fluid motion but instead a loud thunk and then paralysis—that, out of pure irritation from having to reverse-engineer Stäubli's gobbledygook, he inserted this comment into the code:

PLEASE FOR THE LOVE OF WHATEVER DEITY YOU WISH TO USE HERE DO NOT, I REPEAT, DO NOT EVER REMOVE THIS WAITENDMOVE () // OTHERWISE BAD THINGS WILL HAPPEN TO ALTER () AND, THEREFORE, WILL HAPPEN TO YOU!

Chris Johnson also noticed something funny in the arm's wrist as it moved to its standby position, beside the end of the conveyer, which he termed "home." The motion was not as fast as it should have been, and the wrist's fluidity, it dawned on him, seemed temperature-dependent. Repeatedly, Stäubli insisted there was no problem, that such a theory was absurd. Eventually, Stäubli sent a technician to Victor, saw what Chris Johnson had been describing, and admitted that the grease in the arm's wrist joint appeared to be stiffening in the cold. The grease at the arm's wrist was thicker than elsewhere, on account of the planetary gear within. "They've told me I've been wrong a lot," Chris Johnson said. "I've been wrong a couple of times."

Like Scott Peters and John Nolan, Chris Johnson was a believer in questioning everything—not to be an ass but because he wanted answers. "If someone asks me to do something their way, and I wanna do it my way," he said, "I'll do it their way and show 'em it sucks. That's the easiest way. Then we'll do it my way." He also said, "If you just hire people who don't challenge conventions, your company will fail."

You could see why Scott had hired Chris Johnson. It didn't hurt that at home, he had built an Arduino-powered home-brew-dispensing robot. It grabbed a glass, poured a beer, and alerted him via his iPhone. He had five taps. Often, beer in hand, he'd watch *Mr. Robot* and wonder if life got any better.

11.

A Stage to Stand On

Whenever Scott was out of the office, Rocky got crafty. He staged photos of SAM playing Ping-Pong, or of his engineering colleagues passed out, feet up, peanuts scattered everywhere. Then he emailed the photos to Scott—who was off pitching SAM to the big boys of the construction world—and said that some important document was attached. Rocky was feeling good that June, as SAM's new frame started getting built at PMD. There was still no shortage of stress, but it was different. Where the stress at Kodak had been painful, the stress at CR was beautiful, because it arose from fun challenges.

Things seemed to be moving sufficiently enough in the right direction that Rocky decided that it was time for him to go. He was sixty-three, his mustache graying. Before going, he told Scott that building SAM had been one of the highlights of his career.

Three months later, Erwin left, too. Like Rocky, he'd committed to getting a machine in working condition. Now that everything seemed to be coming together, his job was done.

With Erwin's departure, John Nolan at last got a title. He was CR's systems engineer, even if Erwin teased him by calling him "Erwin 2.0." Somehow, though, Tim inherited Erwin's hard hat.

By then, they had an entirely new machine. Whereas the engineers had put the Alpha machine on rails, they put the Beta on a simple low-riding chassis with four thick red rubber wheels. As if the infant machine required training wheels, it also had four smaller wheels, each on a hinged arm—whose purpose was to sense that the three-thousand-pound assemblage had reached the edge of the scaffold, and automatically stop it from careening off the end. On the machine's front—the wall-facing side—they mounted the smaller red Stäubli left of center (a position that made it a left arm). Low and to the Stäubli's right, they placed the black plastic mortar nozzle. It stuck out like a muffler pointed the wrong way.

Where the Alpha had two hinged brick conveyers that protruded from the left side of the machine, the Beta had just a single level conveyer that jutted out from the right side like a diving board. Where the Alpha had only a small, mixerless cone-shaped hopper at shoulder height, the Beta had a large square hopper, right in the middle of the back, at waist height, to make shoveling mud into it that much easier. It looked like a beat-up utility sink, except that it was covered with a sturdy steel grate. To the hopper's right, the engineers placed a generator, which looked not unlike a shiny white washing machine, and above it they secured two propane tanks. To the hopper's left, they placed a stainless-steel cabinet of circuitry topped by a control console with a dozen buttons, knobs, and switches, which, in combination, hinted more at an airplane cockpit than a car dashboard. Above that, on the back left, a trilight resembling what was found atop sailboat masts was mounted. It ran green when the machine was running, yellow to indicate a warning, and red if there was a problem. It emitted two alarms—a cacophonous *chirpchirpchirpchirpchirp* if somebody was blocking the laser, and a much gentler *tweet-tweet* if the machine didn't like a particular brick.

Between the front and the back of the machine, the engineers placed a Plexiglas windshield—not for speed but to keep mortar from bespeckling their delicate optical equipment as well as the Stäubli. And on both sides, they placed touch-sensitive rubber bumpers, so that the machine would not inadvertently roll over anyone or anything. To be double sure, they put big red STOP buttons on both sides as well. The big buttons presented somewhat alarmingly but paled in comparison to the sticker on the electrical cabinet that said DANGER 240 VOLTS.

In addition to the new machine, the engineers had a pair of poles (one "dumb," one "smart") and a track of such gleaming stainless steel that it posed a serious slipping hazard when covered in morning dew.

They discovered this soon after dragging their Hydro-Mobile out back and turning the whole east side of the building into a test wall. The engineers deemed it the danger track.

This terminology they did not mention to TUV, the firm CR hired to conduct a safety assessment of the new SAM. CR hired TUV because a safety-approval stamp seemed likely to make SAM more digestible. TUV, which started inspecting steam boilers in the 1870s, tested solar panels and batteries and toys and couches and cars, but the firm had never assessed a bricklaying robot. It had suggested installing bumper bars on SAM, as well as a grounded copper tether. Scott liked the first suggestion but not the second, so he asked TUV to treat SAM like a car. Cars don't need wires dragging behind them on the ground.

All of this design—in machine, and track, and poles, and laser box—left Tim so occupied with SolidWorks that Scott hired another designer. His name was Kim Heng.

Kim had also studied at the University of Rochester and had graduated a few months earlier. He'd been offered a full ride to Bates College, but Bates didn't have an engineering program. During his time at U of R, Kim worked on the second-largest laser in the world, scurrying around in a Tyvek bunny suit. He'd also stocked groceries and didn't want to do that anymore. He certainly didn't want to make staples on a factory line, like his dad.

Kim ended up beside Tim, on the other side of a cubicle wall from Glenn. He was, as promised, good at design—and it helped that he didn't mind picking up the office phone, so the guys put it by his desk. He redesigned the laser box, rendering Rocky's "one-pound bag" even smaller and lighter. He stuffed it not with ten pounds of shit but with off-the-shelf parts, and powered it with a rechargeable DeWalt battery available at any Home Depot. The battery snapped right into the bottom, as on a cordless drill. Then he redesigned SAM's gripper, eliminating a swiveling finger. In time, in the style of Bruce Schena, he would design thirteen more versions of the gripper. But first, he helped Tim with the track.

The track was less simple than it appeared. Made of heavy-gauge steel, it was strong enough to support SAM, keep SAM aligned in its lane (thanks to a narrow gap), and support brackets that slid out from under the Hydro-Mobile and undergirded long wooden planks where masons spent most of their time. Those narrow sunken planks were an airy but crucial work area from which the rest of the steel deck was at knee height. A mason standing on the planks never had to bend over to grab bricks or mortar. The track, made in sections, had to bolt on to a Hydro-Mobile scaffold seamlessly, so that masons wouldn't trip on sharp edges, and so that SAM wouldn't get stuck like it had on the old roller-coaster rails.

Kim generally spent a third of his time downstairs in the shop, and after a brick nearly landed on his foot, he decided to wear boots to work every day. But at least he didn't fall off the scaffold, like Chris Johnson, who had actually done so twice. The first time, he was working out back with Kerry, and when Kerry looked up, all of a sudden Chris Johnson was gone. He landed on his feet and was fine. The second time, it happened indoors, in the company of John Nolan. Chris was so immersed in the motion of SAM's arm that, as he scooched backward, he moonwalked right off the planks. In both cases, the Hydro-Mobile's guardrails had been removed, because they were interfering with robotic research. Chris Johnson again landed on his feet and was fine. Maybe those protein shakes paid off.

Every week, bricklayers worldwide fall to their deaths from perfectly good scaffolds, or get thrown from faulty or improperly configured scaffolds, or crushed by scaffolds being assembled or disassembled. Masons die when they slip trying to climb from a platform into a window, when planks break, when guardrails give way or simply aren't there; but also when scaffolds are overloaded, not balanced or configured properly, or when they strike power lines. A general rule holds that more time on a scaffold and more trips up and down one present more risk. As such, mast-climbing work platforms like Hydro-Mobiles are safer than traditional scaffolds—but when they fail, disasters ensue. In 2006, a crew disassembling a mast-climbing scaffold on Boston's Emerson College goofed up. A foreman and laborer were killed in the collapse, as was a young doctor driving by on Boylston Street. In just one week in the spring of 2015, three construction workers died in Raleigh, North

Carolina, as did two bricklayers in Toronto. Between 1990 and 2010, a dozen mast-climbing scaffold incidents in the U.S. yielded eighteen fatalities. Worldwide, five dozen men were killed in mast-climbing scaffold accidents in 2013, and five dozen more were killed in 2014.

Until World War II, accessing the face of a building under construction either required erecting a lattice of planks and poles made of steel or timber or bamboo, or hanging from long suspended swings. Suspension setups required rigging knowledge beyond mere square knots, to the clove hitch, scaffold hitch, blackburn, and becket—so most scaffold men were merchant marines. On the Empire State Building, lengthy ropes were so heavy that they were impossible to pull, no matter what kind of pulleys they ran through. In 1947, a Seattle shipbuilder created the first electric scaffold, which was little more than a suspended platform with an electric winch. Not long after, lightweight scaffolds that could be raised by a hand crank or foot pump were devised.

Mast-climbing platforms date back to 1956, when a Swedish company called Alimak schemed up a way to electrify a hand-cranked scaffold. Alimak used a rack-and-pinion drive to raise a steel deck between two vertical towers, the same notion PMD schemed up fifty years later. More mast-climbing platform companies popped up in the 1960s; many ran out of money, changed names, and struggled. However, none was imported to North America until 1982, which shows not only the resistance of the labor unions but also the speed at which innovation takes hold in America's construction industry.

Five more years passed before any mast-climbing platforms were built in North America. They were designed by Andres St. Germaine, a Canadian inventor who'd been approached by masons tired of getting by with apparatuses designed for painters. He devised a wide steel deck that climbed up beefy towers with hydraulic ratchets. (The towers, made of square sections four feet tall, could be stacked to near-unlimited height.) The company that made it was called Hydro-Mobile. St. Germaine ran out of money, so a manager named John Robillard took over, filed for bankruptcy, and moved to a suburb of Montreal in 1994.

In a different Montreal suburb, St. Germaine started a new company, based on a single-mast design, in which the steel deck extended symmetrically from the tower, like branches sticking out from opposite sides of a tree trunk. He called it Fraco, short for France Company, after his wife's surname. A masonry contractor named Armand Rainville liked using St. Germaine's platforms so much that he bought the single-mast design and ran with it. Alleging patent infringement, Hydro-Mobile sued, but Fraco prevailed.

At Hydro-Mobile, Robillard and his partner, Vincent Dequoy, disagreed about the path forward, and when Robillard tried to buy out his partner, the move backfired, and Robillard was pushed out of his own company—so he went and started the third mast-climbing platform company in North America, Bennu. Hydro-Mobile yet again sued for patent infringement. In 2003, Robillard died of a heart attack likely brought on by stress. Bennu platforms are still around, but the company is no longer one of the big players.

Two more companies began to make mast-climbing platforms, in part because the 1996 Olympic Games in Atlanta created such an opportunity. EZ Scaffold, which was still making 1950s-style crank-up platforms,* copied Hydro-Mobile's design but tweaked it to ascend faster.

In 2009, a divorce led to the creation of the fifth big player in the industry.† Premier Scaffolds spun off from Hydro-Mobiles's biggest dealership, after Hydro-Mobile raised its prices and was unwilling to modify its platform design as customers saw fit. Premier's scaffolds look a great deal like Hydro-Mobiles but have bigger ground pads, and simple pinned connectors.

Mast-climbing work platforms—or MCWPs—are popular on big jobs in parts of the country where labor costs are high, i.e., up north‡ and not in Florida or Texas, because while they save a great deal of time, they cost as much as a nice pickup truck. Consequently, in the country's

*The coolest modern crank-up scaffold is made by ReechCraft of North Dakota. Its aluminum platform weighs so little that it may be raised (on lightweight aluminum towers) with two electric drills running at the same time. Because the system is so light and fast and unobtrusive, it was carried *through* the White House for touch-up caulk work on the West Wing. It's not beefy enough for large masonry projects, though.

†Others: Klimer imitated Fraco's designs, and Dunlop just imported European mast climbers.

‡On the huge medical center in Fayetteville that first got Nate thinking about a robot that laid bricks, the masons relied on Fraco scaffolds, but not the modern electric ones. They were crank-up models, rising three quarters of an inch for every crank, which inhibited the pace of bricklaying.

northern cities, the MCWP companies compete for turf. Hydro-Mobile dominates the market with about eight thousand units, but after fifteen years, many of them are nearing the ends of their lives and are covered in rust. As such, Fraco and Premier, who make galvanized units, are invading, looking for masonry contractors updating their fleets of scaffolds. Fraco has about three thousand units across the country but also makes custom scaffolds. During the recession, the company focused on this market—making a triple-masted freestanding ninety-foot-tall scaffold for Chernobyl, and a double-decker scaffold so that crews in Beijing could finish the Olympic pool on time. Fraco made a wild self-leveling scaffold for an overhanging cooling tower in Jacksonville, Florida. Fraco was not around when St. Louis's Gateway Arch was built, but if any company could have built a scaffold for Eero Saarinen's wacky structure, Fraco was it. Now Fraco has broken the Swedish hold on hoists and become the only manufacturer around of MCWPs and hoists and lifts.*

OSHA, the Occupational Safety and Health Administration, does not approve or endorse scaffolds; it just requires that scaffolds meet criteria—strength and load ratings, etc.—and that trained crews set them up properly by, for example, installing guards on both ends. All told, the agency's regulations amount to forty-four pages plus appendices. Though the majority of bricklayers think mast-climbing work platforms are the greatest thing since, well, the trowel, OSHA knows that fifteen people die every day from job-related accidents—and that working at heights presents a serious risk. As such, scaffold citations are some of the most common doled out by OSHA inspectors, and in 2009, scaffolds were their number one most cited problem.

Up on their own private Hydro-Mobile, Glenn and John struggled to get the new machine through a full sequence: pick, butter, place. Summer had turned to fall, and the machine had undergone its own transition. The new machine had a shelf, a Plexiglas window, two propane

*The only reason SAM isn't on a Fraco scaffold is because Fraco, ever the customizer, wanted one hundred thousand dollars to develop a SAM-friendly track. Hydro-Mobile, sensing a chance to steal some of Fraco's thunder, said it would be happy to help design a track for the robotic bricklayer of the future, gratis.

tanks, and a low roof no higher than a person. It weighed half what the old beast weighed. It looked, frankly, like a hot dog cart.

As per the year before, efforts proceeded long past five p.m. and into overtime territory and then the wee hours—so Scott drove home, had dinner with Torrey, and put his four-year-old son and two-year-old daughter to bed, reading to them out loud. As usual, he chose *Goodnight, Goodnight, Construction Site*, whose glorification of construction machinery—a mobile crane, a dump truck, a cement mixer, a bulldozer, and an excavator—reliably pleased parents as well as children. The book featured no bricklaying robot, but maybe Scott's dream, manifest by so much huffing and puffing, could change that.

These big, big trucks, so tough and loud,
They work so hard, so rough, and proud.
Tomorrow is another day,
Another chance to work and play.
Turn off your engines, stop your tracks,
Relax your wheels, your stacks, and backs.
No more huffing and puffing, team:
It's time to rest your heads and dream.

Once his kids were asleep, he returned to the office with beers.

Downstairs in the shop, everyone could tell Beta was much faster than Alpha. Chris Johnson, who knew that the Stäubli's joints needed just four milliseconds to respond to commands, had smoothed out the arm's motion, and Mike Oklevitch was tracking cycle time with a stopwatch. When they got through a full cycle, everyone cheered. The cycle had taken eighteen seconds.

They forked SAM off the Hydro-Mobile, dragged the Hydro-Mobile out back, and forked SAM back onto it—so that now SAM, loaded with real mortar, was addressing a real two-story wall. They built and tore down and built and tore down and built and tore down and filled up many Dumpsters with cement-covered bricks.

SAM's movements were nearly elegant: It placed a brick, rolled eight inches to the right, then reached out and placed another. Actually, SAM started rolling to the right as soon as the gripper let go of the first brick and the arm began retracting. In the few seconds it took SAM to roll a brick width to the right, it had already conveyed a brick to the home position, grabbed it, and begun bringing it toward the nozzle. Slicker still, as SAM worked its way to the right, the Stäubli made its final approach to the wall down and left. Conversely, on the next course, as SAM worked its way to the left, the Stäubli made its final approach down and to the right—so that the most recent brick was always pushed snug against those that were already laid firm.

When snow fell, the engineers erected a plastic enclosure over the wall and the Hydro-Mobile, and it was so airtight that John nearly passed out from the accumulating fumes spat out by SAM's generator. Using two-by-fours, they screwed in some fake windows and doors, then input the measurements of these openings on a new wall map, fed the wall map to SAM, and watched SAM build around them. They even produced a safety manual and laminated it.

In late November, a storm ripped the plastic enclosure—at which point the guys loaded their new bricklaying machine into a snowmobile trailer and, behind a rented Dodge pickup, pulled it to their first real job.

But Scott, who had suffered a tragedy, was not among them.

12.

The General

In fifth grade, Scott Lawrence Peters invented a contact-lens remover. In sixth grade, he came up with earmuff headphones, an invention deemed the best submitted by any elementary school kid in western New York. His mom worked as an occupational therapist, so he was encouraged to think about other people's needs, and also to do crafts.

In seventh grade, Scott started swimming. The idea came from his older sister, Kelly, who had joined the school team as a diver. On a whim, she had tried moving through the water and broken a record at her first meet. She ended up getting recruited to the University of Rochester, where she swam everything, and went to nationals every year in multiple events. At one point, every university record, including diving, belonged to her except one: the 500 freestyle. Distance was not her thing. It was her brother's.

Scott had tried baseball, football, basketball, soccer, track, even golf, but swimming stuck. He did well enough in seventh grade that by eighth grade he was on two teams: a summer league and the varsity team. That year, using a symmetrical stroke a far cry from the lopsided one employed by Michael Phelps, he won the 500 free for the first time, and broke six minutes.

Swimming is a funny sport: You can see your competition clearly at the start, and then suddenly, they're gone and you're all alone. In many ways, it's the purest of sports, more so than running, because you're basically naked and so immersed in a medium that you can't smell or hear anything, not even cheering, and trying to follow the progress of your competitors in underwater glances only slows you down. All you can do is gulp for air while you settle into a rhythm and focus on the path of each arm, keeping your elbow high, hand low, reaching to full extension, sinking your thumb into the water first, taking an S-shaped pull, and finishing, behind your ass, with a flick of the wrist. In no other sport do competitors proceed so head-down.

In ninth grade, Scott kept swimming freestyle and took up breaststroke, becoming the team's best in the stroke. In tenth and eleventh grades, he started competing in the individual medley. His butterfly and backstroke segments were not the fastest, but he'd recover in breaststroke and then hold on to his lead in the final freestyle. By senior year, he was competing in backstroke, too, which made him a utility swimmer. He was good at everything. With his help, the summer league team, the Tigersharks, won their division four years in a row. Scott's coach with the Tigersharks was his uncle Dave, who suffered from myotonic dystrophy, causing his muscles to get weaker and harder to control. He used to tell Scott, *If you could invent a better zipper, you'd make a lot of money.*

His high school coach set a different example. One morning, Scott and his teammates discovered the pool's filtration system wasn't working. The water was chlorinated, but it had a slimy film on the surface. It looked greasy.

Quit your whining! their coach yelled.

The kids kept whining.

Get in the pool! their coach yelled.

The kids did not get in the pool. They said it was gross.

So their coach, fully clothed, fully sneakered, jumped in—and then the kids did, too.

Their coach was a Buffalo native named Dick Beyer, but everyone just called him Coach. His wife they called Mrs. Coach. Fifty years earlier, when Beyer was in eighth grade, he failed four out of five subjects, but by tenth grade, he was the size of a mason. He went to Syracuse, where he dominated on the wrestling team, co-captained the football team, and helped bring the Orange to their first Orange Bowl. There, after they got walloped by the Crimson Tide, he advised his teammates to learn from their defeat. Syracuse named him the Athlete of the Year.

He'd worked construction, carrying eighty-pound bags of mortar, but figured he'd become a pro wrestler. This was 1954. He sent letters to fifty promoters, and not one wrote back. He decided to make a go of it on his own by impressing his first fifty opponents. Needing a persona, he bought a woman's girdle at a Woolworth's, put it on his head, and cut two holes for his eyes and one for his mouth. He called himself the Destroyer and offered five thousand dollars to anyone who could break his figure-four leg lock.

He fought Baron Gattoni, the Italian behemoth who once lifted three hundred pounds in one hand. He fought Angelo Poffo, who once did six thousand sit-ups in a row. He fought Classy Freddy Blassie, who filed down his teeth, the better to bite. He fought Primo Carnera, the only man to hold the world heavyweight boxing and wrestling titles. He fought Haystack Calhoun, who weighed nearly six hundred pounds, and he fought Shohei Baba, the seven-foot-tall force of nature from Japan. He fought sumo wrestlers, karate masters. He fought Rikidozan, who did not like to let anyone cast a shadow on either him or on his own shadow. He fought Great Togo, and Hercules Cortez, and Magnificent Maurice. He fought former Olympians. He fought a 450-pound Canadian black bear named Victor (he lost). He fought Andre the Giant, then drove with him from Springfield to Chicago and watched him drink forty-two beers and a bottle of vodka along the way, not piss once, and then, on arrival, say, "Let's go drink."

Work took Dick (and his wife and three children) from Buffalo to Los Angeles to Portland and included lots of driving. Typical weeks looked like: Portland, Seattle, Walla Walla, Spokane, Eugene; Fort Worth, Dallas, San Antonio, Corpus Christi, Houston, Beaumont, Waco; Nashville, Indianapolis, Detroit. At one unprofitable match, he said to himself, *This is both the biggest learning experience and the biggest mistake of my career.*

His dad had once told him, *Try to make your best decisions during the most uncertain times.* So he kept at it, all the while wearing the mask everywhere he went. He wore it on flights and wore it all over Japan—at Osaka Castle, Ueno Park, Ginza—where the people were bonkers for wrestling. In the mask, he became such a star in Japan that the match between him and Rikidozan was seen by seventy million people, the largest audience for any program in Japanese history. A decade later, he was the star of the number one show in Japan, *Usawa No Channel*—in a mask. He was on the cover of *TV Guide* (Japan) in a mask. The Destroyer was such an icon, and in so many ads, that citizens in Tokyo couldn't walk three blocks without seeing his likeness on a billboard. He was in ads for Fuji film, Cheetos, Drano, McDonald's. He told his kids they were the Von Trapps of Japan. His son, who called himself Kurt Destroyer, had his own mini-mask. Back in the States, Dick was invited by the U.S. Navy to inspect the U.S.S. *Preble*, at Pearl Harbor. The idea was one destroyer meets another. He talked with the captain, ate with the crew, and wore a mask the whole time. It seems certain he went fishing in his mask. He took wedding photos in his mask. He was once on *The Regis Philbin Show* in a dark suit and white shirt—and his mask. He gave one to Regis.

All told, "the man with the million-dollar legs" fought over eight thousand matches. He lost eight teeth and broke his nose six times—but also once kicked an opponent so hard that he left the imprint of his shoelaces on the other man's chest. He twice won the WWA's world heavyweight title.

He retired as the Destroyer, fought briefly as Doctor X (which involved another mask), and then returned to Akron, New York. The year after Scott was born, Dick started coaching high school wrestling in Buffalo. Six years later, he left the city, and wrestling, and took up teaching elementary school phys ed at the all-inclusive school where Scott was a student. He also took on high school swimming. The last time he'd swum was in 1942, when Roosevelt was president. As he put it to his team, "The only swimming I've ever done was at Bare-Ass Beach in the forties!"

Under Coach, Scott's team competed in New York City and spent two weeks swimming in Japan. Coach funded the whole trip by getting the team to sell bottles of Snapple. On the flight over, Coach came out of the bathroom wearing a mask and kept it on for the duration of the trip. Everybody recognized him.

Coach didn't know as much as others about swimming—he emphasized dry-land exercises and rubber-band drills—but he learned what he could. As Scott put it, Coach "just picked it up and went for it." He was, though, a good motivator and a great promoter, and he showed that it was possible to attack anything, even at retirement age.

Scott's name still stands on the school's record board for a relay race. He was, in his own words, a "studious athlete." He played tuba and took piano lessons, but music didn't really stick. His math teacher noticed a different gift. She told Scott's mother that Scott had a mechanical mind but also had people skills—a combination she rarely saw.

Before he graduated, Scott came up with the idea for carbonated coffee, trailing Red Bull by eight years but beating Monster Energy by seven. Even so, he didn't know engineering was a subject until his junior year, when a hiking buddy heading off to Purdue told him about it. Hearing the word, Scott said, *What's that?* His buddy told him the world ran on engineering; it was not just making things, it was how things happened. He told Scott, who'd never realized that his grandfather was more or less an engineer, to look into it.

Scott's roots, and especially his roots with bricks, go a long way back.

In 1545, one of his earliest known ancestors, John Baldwin, was given a Buckinghamshire manor called Dundridge by Henry VIII, after Henry VIII had the manor's previous owner, the Countess of Salisbury, beheaded. Sir John had been the chief justice of the common pleas for a decade; he once executed a man for a murder that was never committed. He was seventy-five years old when he inherited the gabled brick manor, and he died within months.

Leaving that same manor, Sir John's great-great-grandson Sylvester Baldwin set sail for America in 1638, a generation before fire ravaged London to such a degree that the city forever turned to brick. Sylvester did not survive the trip, but his teenage son John did, and established himself just west of New Haven, Connecticut. Six generations later, Emanuel Baldwin married Nellie Ballard, whose family had roots in the Massachusetts Bay Colony. One hundred years before the Revolutionary War, Nathaniel Ballard fought the Wampanoag under Captain Prentice. On the south side of the Saugus River, where the ferry landed on his property, Ballard Street remains.

Emanuel and Nellie had a son named Lawrence, and Lawrence Baldwin married Pearl Yeomans and had a daughter named Betty. At Corning, in New York's southern tier, Betty Baldwin met a young Peters who went by Bud and whose roots, including Martins, Websters, Meads, Corbetts, Orrs, and Tolles, went back to 1768 in New York and 1570 in Scotland. That branch was from the lowlands of Lochwinnoch. His other roots—Whites, Stewarts, Rutherfords, Armstrongs—were mostly English but were also Irish and German and Dutch, and included an inventor named Ephraim Rosencrantz. Ephraim, who was six-six barefoot, patented the design for an improved gas lamp, had an idea in mind for a dray wagon with brakes, and according to family legend, invented the torpedo, for which he was murdered in Washington, D.C. Betty and Bud married and had Scott's father, Jeff.

In 1969, Jeff Peters went to junior prom with Nancy Finger. They'd gone to middle school together and got married in college. Her roots were half German, and her father, Lawson Lawrence Finger, was a genius who loved water. He had a photographic memory and an astronomical IQ, and when he returned from the Pacific, where he lost two platoons in World War II, he took to calling everyone Charlie and started Buffalo Sheet Metal. Scott, his grandson, he called Charlie.

The other side of Nancy's family was English and Scottish. At least one fought in the Revolutionary War. Other relatives fought in the War of 1812 and the Civil War. One ancestor married Sarah Waldo, making her and Scott very distant cousins of Ralph Waldo Emerson.

In memory of his genius maternal grandfather and his paternal great-grandfather, Scott was given the middle name Lawrence, but it might as well have been Nathaniel—and not for the tailor from Saugus. Nathaniel Holmes, born in Roxbury, Massachusetts, in 1639, was the first American-born of Nancy's ancestors. He was the son of George Holmes, who'd lived just north of London and come over to the Massachusetts Bay Colony in 1637. In addition to working as a fence viewer (he settled land disputes, mostly about cattle), Nathaniel was a representative at the General Court, a selectman, and a bricklayer. At age twenty-five, he opposed King Charles's accession, and in 1689, when Edmund Andrews, the crown's authoritarian administrator of New England, was held in captivity, Nathaniel went to Boston to see about establishing a beneficent government that had the consent of the people. He fought the Nipmuc under Captain Hutchinson and contributed to the fund for the purchase of a bell at the First Church of Dorchester. In 1696, putting public interest ahead of private profit, he signed a petition proposing the repeal of the 1692 law that mandated buildings in Boston be made of brick. Because fires posed such a danger, buildings made of timber could be legally demolished, their builders imprisoned. The settlers, though—mostly poor tradesmen—found this overly burdensome; they said limited resources already left them so impoverished and distressed that they were near destitute. Rich foreign arrivals could afford brick buildings with slate roofs, but the settlers could not—and they were not about to become indentured to their new wealthy neighbors. If the law was not repealed, they said, they very well might have to pack up and leave town. The law was not repealed; on the contrary, the government soon threatened to impose fines of fifty pounds on builders of timber houses—but Nathaniel did not leave town. In fact, ten years later, between the Paines, the Baxters, and the Allens, his nephew started a brickyard. Just like Scott, he saw value in brick.

At the University of Rochester, Scott pursued engineering from the start. At first he was attracted to environmental engineering, but the school wouldn't let him create his own major. Then he was into mechanical engineering. Eventually, he settled on chemical engineering and, in a month, wrote a thesis, not surprisingly, on water. More precisely, it covered arsenic in Bangladeshi groundwater.

For the Yellowjackets, he kept swimming long-distance free (now the 800) and, rising high out of the water, got his 100 breaststroke time down to 1:02. He was six-three, with an easily caricatured, perfectly triangular Roman nose, a shaved head, and shaved legs to match. At the start of his freshman year, he met Torrey Podkaminer, a sprinter almost as fast as Kelly, who was visiting on a recruitment trip. Scott also joined the water polo team, which counterbalanced the individual, speed-obsessed, meditative nature of the first sport. The result was two practices a day. On the water polo team, Scott was a hole setter, aka center, which left him facing away from the other team's goalie, trying to turn around and score in sneaky or just plain fast ways. With Scott, U of R's water polo team went to nationals his sophomore, junior, and senior years, taking on Division III schools. In the last two years, U of R got second place, and Scott was deemed MVP. His teammates called him the Monster.

Scott joined Sigma Chi, where he was too busy swimming to drink. When he first joined, he was the only swimmer in the house. Swimmers, though, develop a strange confidence, born by mastering a medium meant for fish. They know they can handle most situations. By his senior year, the fraternity was half swimmers, and Scott's nickname, though it might as well have been the Recruiter, became the General. Once, senior year, Scott tried to outdrink Sigma Chi's pledges, on the condition that the drunkest would have to wear a Santa Claus costume the following day. Scott started by slamming a Long Island iced tea—and doesn't remember what happened over the course of the night. He only remembers waking up in a Santa costume.

Scott stayed a fifth year and earned a master's degree. All the while, one of his friends dated Torrey, and he dated one of Torrey's friends, but the two never dated each other.

In Boston, where he landed a job at Intel, Scott started thinking of himself as a process engineer and bounced between bars at Harvard and the aquatic center at MIT, knowing he was playing on turf where a lot of big ideas had been born. He also played water polo on a team called the Last Drop. In one tournament, he beat U of R. When he returned to western New York and got a job at GM, he got into kiteboarding on Lake Erie, and kept swimming—until his future father-in-law got an idea in his head.

One other idea never left him. His uncle Dave had mentioned a better zipper—so Scott had Tim Voorheis draw up some designs. They held little promise. Scott didn't let the idea go, though, and brought up the subject at his parents' Lake Ontario cottage one night, around a campfire. His neighbor said, *Well, I do some 3-D drawing . . .*

Many iterations later, an improved zipper was born. Until then, it took two hands to align the pin at the bottom of one track in the box at the bottom of the other before you could pull a zipper's tab and engage the teeth. Shaky hands made the initial alignment difficult if not impossible. Scott, his mother, and his neighbor came up with a zipper whose end magnetically engaged, making it deployable with one hand. Scott's neighbor's name was David. Scott's mother's name was Nancy. In the summer of 2009, they patented their MagZip under the name DNS Designs. DNS: David, Nancy, Scott. As with bricklaying, Scott looked at something ubiquitous and saw room for improvement. The technology is now licensed to Under Armour, resulting not just in royalties but in a lot of free clothes. From MagZip users he also gets emails thanking him for making their lives a little easier.

And then one November day, a few hours after sundown, Scott got a call from his father. Scott was on his way to a friend's house. *Go home*, his father said, *and call me back*. A few minutes later, Scott's sister called Zak at the office. Zak picked up, because a call from her at that hour meant something must be wrong.

Scott's younger brother, Steven, had killed himself.

Zak raced home and, in silence, gave Scott a ride to Akron, where Scott's sisters also lived. Just before leaving the interstate, Scott told Zak to oversee this and that at the office. Not wanting to intrude on the Peters family, Zak gave Scott a big hug, then returned to his sister's house, to Scott's wife's house.

Scott did not return to the office for two weeks.

There was a funeral in a church.

There was anguish and examination. Steven had been diagnosed with Lyme disease a few months before, perhaps having picked it up on the Appalachian Trail, but nobody knew he was struggling. He was undergoing treatment and had just switched to a new regime, but he told no one of the pain. He was so smart—smarter than Scott, Scott admits—so why did he not ask for help? Only days before, his parents had visited him at his apartment. "I love you, son" were both of their last words to him.

Steven Edward Peters, named for Grandpa Bud, had been at the top of his class and then dropped out of college. He'd worked for his dad for a bit, then gone back to school for massage therapy. He'd been searching for a way to apply himself.

With his older brother, he used to explore the Onondaga escarpment, hunting for fossils and the cave where the counterfeiters of yore allegedly buried money. The brothers used to climb high Adirondack peaks and play competitive euchre. Scott thought of Steven every time he sailed, not just because of the time when the two of them, desperate to reach the shore of Lake Ontario before a lightning storm enveloped their tiny craft, had lain down on the bow and paddled half a mile, but because, growing up, their father had organized a race called the Stevie Cup, always on Steven's August birthday. Days before the event, his father would put up signs in the neighborhood, nail Steven's name to trees.

When Scott returned, he was changed, as if a candle had been blown out. He was quieter and less able to concentrate. He was reminded of his brother constantly and fiddled persistently with the thin gold chain he had put around his neck. It was Steven's necklace. In his office, he kept a photo of his brother.

"You don't recover," Scott said. "You just decide to move on."

To move on, he immersed himself in work, becoming more driven than ever. He stopped exercising and, through the fog, devoted even more energy toward SAM. Which was fitting, because SAM was very much a minimally viable product and needed all the help it could get.

PART II

THE REAL WORLD

13.

Vegas

As far as venues went, the parking lot beside the Bricklayer 500 arena was probably the best place in the world for SAM to make its public debut. Las Vegas wasn't a brick town (it had more imitation bricks than real ones), but in every direction, something new glimmered. Las Vegas was a city built on speed—on fast building and fast money—with negligible concern for the traditional. Only there could one live a life that entailed waking up at three p.m., working in shorts, and wearing sunglasses inside. It was a place where anything went, and anything was possible.

At the end of January 2015, nearly the entirety of the little start-up swooped into town. While Scott and Nate and Zak flew across the country, the engineers got a head start in the rented blue pickup and a Penske box truck, into which they strapped SAM. This, of course, was not exactly legit: Lacking a registration number from the U.S. Department of Transportation, they didn't abide by the same rules (documented rest and health) that other commercial haulers did. As usual, they were winging it. On the other hand, the Penske barely held 30mph going over the Rockies, and they took turns driving, so it wasn't like they got away with too much.

After settling into shared rooms at the Riviera, the engineers spent a week setting up CR's demo area, which was not insubstantial. First they erected a sixteen-foot wall, just like the real thing, of metal studs and gray sheathing and a base of gray cinder blocks. In front of that, they positioned a rented Hydro-Mobile scaffold, having bolted their track to its sixty-foot deck. Behind the scaffold, they assembled two large red tents, complete with tables and chairs and brochure racks, and around all of it—the wall, the scaffold, and the tents—they set up a barricade of the variety used to cordon off marathons. From the Hydro-Mobile's thirty-foot towers, they even hung a huge Construction Robotics banner.

Onto opposite sides of the scaffold, as counterweights to the Volkswagen-heavy SAM, they forked two cubes of bricks—each cube weighing a ton. Onto the middle of the deck, they forked SAM and another cube of bricks. And then, as World of Concrete officially began, they forked up a tub of mortar—not Mortar Mike's play mortar but the real thing—and started laying for the crowd.

As the convention unfolded, the engineers babied their machine and tried to look cool and calm as they built upward. Aside from Scott, who sported khakis, they all wore jeans and gray golf shirts with the company name on the chest. John took SAM's controls, mostly; Mike Oklevitch and Chris Johnson generally shoveled mud into SAM's hopper and loaded bricks onto SAM's conveyer. With an elongated wrench called a brick tong, they grabbed ten at a time from the bricks stacked up in the cube, then plunked them down on the rubberized belt of the conveyor.

All the while, the men shook a thousand hands and explained how the heck the machine worked. Lasers got the most attention: not just the positioning laser fired from the box on the smart pole, but the many smaller lasers on board SAM, at the end of the conveyer belt. These precisely measured the dimensions of every brick SAM touched, so that SAM could center each one in the wall appropriately. The engineers explained the meaning of SAM's various alerts—the chirping, the blinking lights—and showed how straightforward it was to keep SAM supplied with bricks and mud.

The engineers pointed onlookers to the nozzle, which pointed, suggestively, straight up, and they emphasized SAM's exceptional mortaring—neither insufficient nor excessive. A visitor with the inclination to look closely could see, indeed, the smooth seal of mortar on the leading edge of every brick SAM buttered. As those bricks went down, whether the machine was rolling left or right, they always found dry bricks to dock next to—and left a clean, smooth head waiting for the next slathered brick.

Sometimes the engineers pointed to the machine's most recently added feature, officially called "auto-pumpout" but more prosaically called "poop mode." To prevent mortar from hardening in the augur tube, the machine automatically ejected mortar from its nozzle every few minutes, as if plagued with the runs. To minimize the mess on the planks below, the men hung a white plastic catch bucket beneath the nozzle and tried to empty this "poop bucket" right back into the hopper regularly. Nevertheless, mortar tended to accumulate beyond the lip of the bucket and reach for the sky. As if infected by his kids, Scott called this poop tower a sand castle.

And that, unbeknownst to Scott, was probably Construction Robotics' best attribute: Its engineering and marketing teams were one and the same. Where most other companies at the World of Concrete convention employed teams of slaphappy salesmen who relied on scripted language to push their products, CR—with the exception of Zak and Chris Raddell—employed the designers and builders of SAM to show it off. If they were less than smooth, it was not because they were bullshitting their way through overslick technical-sounding mumbo jumbo; it was because all they knew was technical stuff, and on top of working fourteen hours straight, they'd stayed up late to raise SAM's poles and take measurements so that the machine could begin laying promptly. If tricky questions were asked, applications engineers back at the home office did not need to be summoned; there were no other engineers. If they couldn't answer your questions, nobody could.

Mostly, people just wanted to behold SAM up close and be mesmerized by the Stäubli's fluid, perfect motion. To do so, they climbed a short aluminum ladder and stood beside the machine on the narrow steel deck of the Hydro-Mobile. Some had specific things they wanted to see. One day, to test SAM's dynamic compensation, the owner of a masonry company jumped aggressively on the planks, making the enormous scaffold wobble so much that Scott worried it was going to bring the whole thing down. (He nearly shit a brick.) SAM, amid the ruckus, continued laying bricks as before. On another occasion, John noticed that twenty people had clambered onto the scaffold, and he got so nervous about stability that he rolled SAM to the middle of the deck and turned the machine off. That's what you got with engineers in charge of spin.

But spin was required when it came to one big, oft-lobbed question: Had the machine worked on a real jobsite? Most bricklayers, it seemed, were vaguely familiar with the long history of overhyped bricklaying machines, and knew instinctively that the only litmus test for any piece of equipment was experience on a real construction site.

Indeed, SAM had, barely a month before, been on a real job: a barracks in Fort Lee, Virginia. The owner of the masonry firm that held the contract had read about SAM in the news and eagerly put SAM in front of the five-story building. But SAM hadn't performed very well or stayed very long. It had stayed no longer than the Alpha had stayed on the PMD job. And just as Scott had withheld some information about the PMD job during that first publicity event, he did again about the Fort Lee job.

Scott tried not to mention that at its best, SAM had placed 531 bricks in a day, which was twice as good as the Alpha but barely on par with human performance, or that SAM's average had been more like 300. He didn't mention that bricks did not reliably flip over on the seesaw table at the end of SAM's conveyer, or that many of the bricks that made the trip slipped in SAM's gripper, so they got placed at a slight angle, forming a sawtooth along the top edge of each course. He didn't mention that the laser box, and hence the laser line, rattled in the wind, or that adjusting the tabs on the poles took forever. He didn't mention that sometimes SAM took a minute or two to place a brick, or that measuring each section could take half an hour if someone blocked the laser. He didn't mention that he'd found it hard to maintain momentum in a wall broken up by so many windows, or that SAM ran for only a fraction of every day. He especially didn't mention that SAM never made it to the fifth floor and barely made it to the fourth.

He didn't go into the profound, universal frustration of the job itself. Chris Johnson had gotten so flustered with a bug in the robot code that at one point he'd shut SAM off for twenty minutes and said, *Lemme think!* Not used to working around SAM, the masons on either side of the machine kept blocking the laser, prompting the machine to emit a squawk that annoyed the men more than it encouraged them to reposition their ample midsections. After only ten days of working with the machine, the crew had enough, and one of CR's own engineers had too much, too. Paul, the Harley rebuilder, resigned, telling Scott it was all too much exasperation for not enough pay.

And Scott certainly didn't mention the impression that a real construction site had left on his employees. The cold had been so brutal that the men had cozied up to the exhaust from SAM's generator; the wind so strong that no number of hoodies could block it; the unabating smell of porta-potties so rank that it induced gagging. To Scott, the most difficult aspect was all of the delays; he hated waiting for other men to deliver tubs of mortar, for masons to cut bricks, and for masons to place the half-bricks at the ends of each course, where SAM's arm could not reach. And then there was the mud—not the mortar but the earthen concoction that slathered the whole site to an astonishing depth. Mud so covered CR's engineers and their rented truck, and quickly enough, their hotel—not just their shared rooms but the hallways—that hotel staff began to leave notes on their dressers that said, *You're making too much of a mess. If you don't stop, we're going to ask you to leave.* Beside the notes were pairs of hospital booties. SAM, meanwhile, looked like it had fought with a bayonet at Bull Run.

Scott, per habit, had pushed too hard, too early, long before his company or his product was ready. The result was a job, yet again, that registered as "incredibly painful." But the alternative was worse. Scott knew that SAM needed to get out there, not just so he could say, *Yes, in fact, SAM has been on a real job*, but so he and his engineers could learn from the effort. As Coach had advised: *Get in the water.* They had dived in deep and sunk.

But already, they were learning how to swim—and that was what Scott focused on.

The competition was staged just west of SAM and drew spectators hours before it began. Fans stationed themselves in bleachers, which rapidly filled to capacity. Along the sidelines, more spectators huddled in groups in bright matching shirts. Some carried homemade banners. Around SAM, the much smaller crowd dissipated.

It was a clear, warm Wednesday, and a passerby would have been forgiven for suspecting a high school football game was about to unfold. By noon, the emcee was entertaining a crowd of about four thousand. "This is our Super Bowl!" the emcee roared over loudspeakers. He told fans to ready themselves for "complete madness."

The competitors certainly looked like football players. That or Navy SEALs. They included a former professional motocross rider, a taut three-time state high school wrestling champion, and a pair of very fit, very tanklike Utah brothers whose Mormonism obviated the need to forgo any unhealthy substances. One linebacker-size competitor had lost thirty-five pounds since qualifying, primarily by cutting back his daily Heineken intake from twelve bottles to two. Most had spent weekends building practice walls and, as such, appeared more bulked up than usual. (Even Chris Johnson, the bemuscled weight-lifting engineer, paled in comparison.) Many had given up cigarettes or put in two and a half hours daily at the gym and watched videos of former competitors and of themselves.* Among the twenty-four contenders in the Bricklayer 500, there were nine previous podium finishers, including the record-holding two-time champion bricklaying phenom Garrett Hood.

The national anthem was sung, and four thousand baseball hats were removed. Up on the scaffold, SAM did not pause; nor did the engineers. Observed or not, they had a machine to attend to, a wall to build, a big number to hit. *Bricks in the wall*, Scott kept muttering.

And then, with a countdown, the Bricklayer 500 began. With their sights on eight hundred bricks, the competitors built the ends of their walls first, then slathered a whole course in mud and got into a rhythm, placing a brick every few seconds. The competitors used one bare hand to place bricks and a trowel-wielding hand to manipulate mortar. In a curling upward swipe not unlike a canoe stroke, they scraped the mud

*Legend has it that when one competitor went to watch a videotape, he discovered that his wife, with the help of his best friend, had used it to record herself in flagrante delicto. Apparently, she'd figured the last thing her husband would watch was a video of himself building a brick wall.

that emerged from the bed onto the head. It was about this fast: Down, wipe, head. Down, wipe, head.

Meanwhile, SAM kept laying, just as consistently. The machine's methods and motions were different, but the unbroken rhythm was familiar.

As a frantic hour proceeded in the parking lot, a quiet one unfolded on the Hydro-Mobile, barely fifty yards away. Freed from attending to visitors and answering questions and playing defense, the engineers grabbed lunch, rested their vocal cords, reapplied sunscreen, stretched.

When, with another countdown, the hour ended, only thirty-year-old Garrett Hood looked unscathed. The rest of the competitors were flushed and sweaty and sunburned where they weren't covered in mortar. One of those masons, a Tennesseean named Fred Campbell, had placed twenty more bricks than Garrett Hood, and incurred no deductions, even if his bed joints were uniformly thick. He'd placed 743.

Scott, the distance swimmer, was unimpressed by Campbell's first-place sprint. The whole point was to lay bricks quickly without exhaustion.

But the crowds loved it. Under the Jumbotron, Campbell hopped in his new truck, and the emcee jumped up in the bed and then invited Campbell's contingent of friends and family, all in matching orange shirts, to pile in. They could barely contain themselves. Campbell, who had won a truck just two years before, gave this one to his tender, since it was his tender who had encouraged him to compete.

Fans leaving the BL500 arena passed SAM on their way back to the convention hall, but after this pinnacle performance by man, the machine seemed less than impressive. SAM wasn't a third as fast as the world's fastest bricklayers. Eighteen seconds a brick? That was nothing to tweet about. Since many of the passersby had a few drinks in them, disparaging remarks emerged in greater propensity, but the CR guys did not engage or show their irritation. Instinctively, they knew better than to inflame the ire of bricklayers. They knew the attitude they were up against.

By the end of SAM's four-day debut, though, they had plenty to boast about. SAM was officially declared the convention's most innovative product—and the machine was still evolving. On one day in Las

Vegas, SAM laid 1,137 bricks, which was more than most bricklayers regularly laid, even if it was nothing any of the BL500 competitors would have boasted about. And this progress had come in the wake of just one real job!

Little did the men know that despite this small victory, it would take a dozen jobs, over the next eight seasons, before the industry found SAM worthy.

14.

Laramie

Four months passed before CR got SAM out in the real world again, on a real job. In that time, the little company prepared, and everyone took on a role. For the engineers, that meant eliminating the seesaw table at the end of the conveyer belt and developing two updates of SAM's software, both aimed less at speed than at eliminating bugs and glitches. Meanwhile, Zak set his sights on buying a custom trailer and a real work truck (and getting a DOT number and medical card). He chose a black Chevy Silverado 3500, a dually strong enough to tow SAM and a bunch of extra gear, and capable of carrying a bedful of heavy Hydro-Mobile track parts. It was not unlike the cushy specimens driven by Bob Boll and Garrett Hood. Scott's focus was on hiring another engineer, who, as luck had it, walked into the office a week after World of Concrete, having seen an ad at Rocky's alma mater.

Like Rocky, Stephen Kean had studied electrical engineering and had experience building machines of his own design. At RIT, he'd written a master's thesis on the efficiency of particular wind turbines and built an eight-foot model. Rather than join a fraternity, he'd interned for a start-up called Environmental Energy Technologies. EET made a novel exhaust-cleaning system that used forty thousand volts to produce a corona, which in turn produced oxygen, which then bound to pollutants coming out of tailpipes. To test the device, EET drove an old diesel shuttle bus around town, piping the exhaust into the cabin and through the gadget before sending it out into the world. The shuttle bus was so old it produced too much exhaust—and all those volts created so much static that soot clung to every part of the device. Stephen was never sure if the technology worked, or if the head of the company knew what he was doing, but it was of little consequence, because at the end of 2014, EET ran out of money—so Stephen began hunting for another job and stopped messing with the corona.

He didn't, however, stop messing with cars. Before RIT, he'd spent two years wrenching on transmissions at AAMCO and, ever since, could be found tinkering with half a dozen jalopies that he owned. Stephen did not think of himself as creative but did identify as hungry to learn—and this was evident in his tenacity. When he bought an old house and began renovating it, he found the work trickier than the automotive variety but taught himself how to proceed. When he took up break dancing, he learned to walk on his hands, and when he took up swing dancing, he got good enough to compete. Thirty-one years old, he was precise and attentive and talented and, down to his trim V-necks and crisply parted dark hair, looked like a mini Clark Kent. Scott hired him as the company's first field-service technician, and it's doubtful a more qualified candidate existed anywhere. Stephen's job would be to operate SAM wherever the machine went. In other words: everywhere.

Throughout the winter and spring of 2015, Scott pushed Stephen to build, build, build—not just to get him up to speed and comfortable running SAM but to let the other engineers address whatever bugs he found. For days on end, Stephen ran SAM with a clipboard at his side, noting every little machine hiccup. The pages on the clipboard made their way upstairs, where Scott hurried his engineers to attack the various issues with great urgency, before their next job began. That job was to be far away—in Laramie, Wyoming—and certain to be trickier than normal (not that they had a "normal" standard yet) for a couple of reasons. The first reason was that Soderberg Masonry neither owned nor used Hydro-Mobile scaffolds, even though Soderberg was one of the top three masonry firms in Colorado. Soderberg owned an ungodly amount of frame-and-brace scaffolding—enough to clad a thirty-story high-rise in the stuff. Of course, assembling that much scaffolding took a month, but Soderberg's crews approached such work patiently, because they were used to it. They were not at all used to Hydro-Mobiles.

The second reason was that Laramie might have been the toughest place to build in the country. High and dry, it was also windy, either frigid or scorching, even sooty—and notoriously tough crews had a history of making jobs there miserable. Conditions just an hour north of Fort Collins were so bad that most companies avoided bidding on jobs there, and companies that did raised their bids by as much as 50 percent so as not to come out in the red. Add to that the pressure of a building project with a hard opening date (the new high school had to open on the first day of school), punishable by daily five-figure fines, and it was hard not to see a cauldron awaiting. But Chris Raddell had met one of Soderberg's owners in Las Vegas, and this man, though he anticipated trouble, said SAM was a curiosity he wanted to see. The last reason, though, was the simplest: Scott's wife was about to give birth to a third child, and Scott knew his attention to detail was about to suffer. His staff would be, if not on their own, at least more isolated.

So Scott found great relief as Stephen quickly figured out what it took to manage SAM and set a new daily record just shy of twelve hundred bricks.

From the start, the job had bad juju. Months before Zak, John, and Stephen showed up in Laramie towing SAM, a man grading a bank of earth on the west side of the new high school had been injured when his bulldozer overturned. Then, on the morning of CR's second day, two steelworkers fell thirty feet and sustained serious injuries. OSHA was rumored to be inbound. On that same June morning, Zak, John, and Stephen were attending to SAM twenty feet up on a rented Hydro-Mobile when, all of a sudden, buffeted by Wyoming's incessant wind, the engineers got the jitters. The air was dry, the mud was thin, and the block wall in front of them wasn't braced. At least nobody was sure.

Because the Hydro-Mobile weighed even more than SAM (and carried the weight not just of masons but also of bricks and mortar, a tub of which weighs half a ton), it had to be braced to something sturdy—like the frame of a building—as it ascended above the first floor. In Vegas and in Victor, where they'd operated below sixteen feet, they'd gotten away without such bracing and never felt a seasick swaying. In Virginia, the scaffold's towers had been fastened to the solid framework of the barracks. Now John and Stephen began to doubt that the block wall they were tying in to qualified as sturdy. With roof trusses, it would have been fine, but the installation of those trusses was on hold following the steelworkers' accident. So went the choreography that is construction: One changed note altered the whole tune.

Zak, knowing his brother-in-law wanted bricks in the wall, contended that the wall was freestanding, and checked with the foreman, who agreed. But Stephen and John disputed the claim, especially since neither Zak nor the foreman was an engineer. They knew the wall was stiff—it was built to withstand eighty-mile-an-hour winds—but without steel beams running across the roof, or some form of buttressing, it was little more than a bulked-up lemonade sign. Rather than risk pulling the whole wall down, they wanted to retreat and get the facts from a structural engineer. In the meantime, robotic bricklaying came to a standstill.

From the dusty trailer, the engineers called Chris Raddell, who had the building's plans. The plans did not yield the desired facts. Stephen, as eager to get bricks in the wall as Zak, wondered if they could work on a different patch of wall, but nothing else was ready. John summed it up: The only thing that was ready wasn't really ready. Or they couldn't guarantee that it was.

Fed up with the insistently logical engineers, Zak called the company from whom they'd rented the Hydro-Mobile. The company's rep advised talking to a structural engineer. This exasperated Zak so much that he got up, left the trailer, climbed into the cab of the new truck, and called his father.

Nate might have been impatient, but he was not imprudent. His opinion overlapped with that of the engineers, frustrating Zak even more. SAM was two thousand miles from home, under a bright blue sky, and bricks were not going anywhere. Instead, team CR bickered the day away.

The bickering resumed the next day and soon rose to near mutiny. Zak wanted to get going and promptly resume bricklaying, in part because he knew that the foreman, who claimed he could lay two thousand bricks a day, was unimpressed with SAM. Stephen and John, though, refused to even ascend the Hydro-Mobile. Zak and John—who had grated on each other in the confines of the office—nearly tore at each other's faces under the real-world pressure. Zak, who figured he had the most experience on construction sites, was pushing like his dad to get going. John, who figured Purdue had given him an engineering degree for a reason, slammed on the brakes—insisting on the official, credentialed, rational word.

Meanwhile, the bricklayers on the frame-and-brace scaffold just around the corner from SAM kept laying bricks the old-fashioned way.

To John, the cause of SAM's idling seemed to be that Construction Robotics, which was designed to sell a product, was now selling a service. It didn't help that Construction Robotics got treated like a subcontractor's subcontractor.

Because it was too windy to make a phone call anywhere else, the three men piled into the Chevy to call Scott. He grasped the matter quickly, calling it "fairly simple," and came down exactly where Stephen and John and Nate had. Zak passed along the foreman's opinion, and another previously unmentioned. Asked whether the wall was sturdy enough, Soderberg's foreman had put it this way: *If your guys aren't comfortable up there, then they're in the wrong trade.* At this, there was a pause. Scott said he'd like to call the project manager at Soderberg.

Still in the truck, they called Soderberg's project manager, and Zak—now a convert—did the talking. "We're used to having a structural engineer sign off on projects," Zak said, "and without those beams in, we just want to confirm that wall can withstand three thousand pounds of tension or compression." Soderberg's man said they usually sought engineers on walls over 125 feet tall, but someone should have documentation. He just wasn't sure he could get it before Monday.

Their week looked shot.

While Scott and Nate engaged in diplomacy over the phone, team SAM and team Soderberg gathered on the ramp of CR's trailer toward the same purpose in person. Both teams wanted to get busy laying but had different expectations. Zak said what needed saying: The Hydro-Mobile had to be properly secured to a sturdy wall. To this, Soderberg's crew chief said, "We built these walls with wind twice as high as this."

CR's guys could have said, *Right, but you didn't build the walls with*

Hydro-Mobiles, and this Hydro-Mobile, with which you have no experience, weighs five tons, and SAM weighs another ton and a half, and all this weight exerts a lot more force on that wall, so we need to be sure the wall can handle it before making a brave attempt.

But another Soderberg mason said, "This fuckin' thing's not going anywhere."

Again, someone could have said, *Well, that's great. We love working on a solid wall. All we need is a certified engineer to sign off on that, which should be easy, and then we can get going.* Instead, Soderberg's crew chief piped up. "I coulda scaffolded and had this done up to the windows already," he said. "I'm just being straight with you guys." Continuing with the straightness, he said he'd been in Laramie for eight months and lost only one week of work to the weather, having worked in wind twice as strong. He said Soderberg had already billed too much overtime, and with the long drive home, neither he nor his guys would be working late.

All the while, Zak kept his hands behind his back. John nearly hid in the trailer's interior. Only Stephen refused to be intimidated by the swagger and posturing. But before everyone headed off to lunch, it was Zak who conceded. He said he just wanted to make progress and was willing to move on to a more suitable wall—one reinforced with a structural steel frame.

The next week picked up where the last one had left them: nowhere. Soderberg reported that the structural engineer they sought did not exist.

Cowed into submission, Stephen and John ascended the Hydro-Mobile, whose deck had been rendered a glorious mess. It was covered in blue foam panels and galvanized brick ties and dozens of the green straps and pieces of cardboard used to protect bricks in their cubes. There were mortar boards scattered about and a bucket of water planted in the middle. Their track, slathered in dried mortar, was a mess, too, so they shoveled it clear and picked up what they could. Where, back in Victor, they'd worked with fastidious engineers under the oversight of a hyper-organized boss, they now were at the mercy of some foreman with leathery skin and a certain attitude.

They untarped SAM, turned the machine on, loaded it with bricks and mortar, and crossed their fingers. But the section of wall was so narrow—under twenty feet wide—that, after only five courses, Soderberg's masons asked if they could turn the machine off and work by hand. What they said was: "SAM is in the way." The last ten courses went in with human paws. One of the masons on the wall to the right, who had taken to calling SAM Samuelito, saw the robot doing nothing and said the *gordo* little robot was *durmiendo.*

At the day's end, Zak announced their new plan: Move to a wall around the corner, lay bricks there for a week, and get out of town. Back in Victor, Scott decided that flying out to Wyoming made no sense. His excuse was his newborn son, but really, throwing money at a money loser was silly. John, Stephen, and Zak were on their own.

Once the deck of the Hydro-Mobile had been lowered, moving the whole contraption a few hundred feet took an hour and some very skillful maneuvering of a telescoping-boom forklift, involving eight-point turns and wheels off muddy ground. Once the scaffold was positioned before the new wall, the guys cheered up for the first time in ten days. To Stephen, the wall before him was fantastic. It was 35 feet wide, 92 courses high. That was 3,200 bricks. Knowing two perfect days might not be sufficient, Stephen called it a three-day wall.

It ended up taking a week and a half. Stephen could never coax more than nine hundred bricks out of SAM in a day, because something always came up. One day, a cement truck parked in front of the Hydro-Mobile, blocking the delivery of mud tubs for nearly the entire morning. Another day, with the delivery route clear, the mortar mixer on-site broke, so there was no mud that could be delivered. At one point, Soderberg was short eight masons, so there was no one to strike SAM's joints. At another, Soderberg swapped out the regular crew, and Stephen was left with what he called their B-team. They had no idea how to work around SAM, which slowed everything down. But by their second day, the new crew figured out life near SAM, and to Stephen, things began to proceed "like Chutes and Ladders."

Still, conditions remained challenging. It was so hot that Zak started stealing hotel washcloths to wipe off his head, and sunburn began to take a toll. Applying sunscreen helped only minimally; their skin resembled greased sandpaper. While other subcontractors could take shelter in air-conditioned modular offices parked around the enormous site, CR had to make do with what was, for all intents and purposes, a horse trailer—small, uninsulated, poorly lit, and full of red dust. Desperate to stay hydrated, Stephen drank so much blue Gatorade that his shit turned green.

Eager to leave, he worked two Saturdays and stayed one night until eight-thirty p.m. Thirty feet up, near the top of the wall, he felt a shimmy in the Hydro-Mobile as SAM rolled along, and he got nervous. But he also loosened up. Unknowingly imitating Rocky, he leaned on SAM and gave it a nudge whenever the laser missed the target on the gripper, helping it along. This technique, of course, appeared nowhere in the manual. As the end of the project neared, he joked with Zak, started to swear as much as anybody else on the site, and—holding a mug of coffee in his right hand while operating SAM with his left—whistled a tune as the bricks went in.

Finally, on June 23, at three p.m., with no mason and no tender, Stephen topped out the wall.

It did not take lengthy analysis to come up with a word describing the job. It was a *disastrophe*. CR made a terrible first impression. There was more waiting than production; more dispute than accord. SAM's performance was equally inept. In twelve workdays, SAM laid sixty-six hundred bricks when the original plan had been ten thousand. SAM ran only a third of the time, and SAM's best day was nine hundred bricks.

Zak, who had warned against attacking narrow walls—he called them "no-wins"—ascribed fault to the masons unwilling to put in extra time to tie in properly and move poles at the end of workdays.

Stephen found fault with the nature of the beast itself: construction. To him, the job was a series of unfortunate events—a bunch of "shenanigans." No day went perfectly because they were at the mercy of fate, subject to personalities and politics. Having wrapped up his first visit to a real jobsite, he predicted, "There's just always gonna be something."

Scott found subjugation to forces beyond his control frustrating. Coming from the computing and automotive industries, he was boggled by construction. It was an unpredictable mess. Known problems, no matter their size, seemed more manageable than unknown ones. As such, SAM never seemed set up for success. "When our system is running, it runs good," he insisted. He wanted more up-front planning so that he could demonstrate what he had.

"One of these times," he said, "we're gonna hit a home run."

15.

Lunenburg

By July 2015, SAM's cycle time was down to thirteen seconds, which amounted to 275bph. Chris Johnson was sure he could drop more time and just as sure that he could deliver twelve hundred bricks in a day. Scott liked the sound of that number and, as an incentive, told his engineers that anyone who compelled SAM to hit that mark would earn a thousand-dollar bonus. Chris Johnson wanted to hit the mark on the first full day on the job in Lunenburg, Massachusetts. It was another school, but that didn't faze him. As the job approached, he flexed his bulging biceps more than usual and kept imagining a new server he'd specced out, or a new snowboard, or the PhantomX hexapod.

Ever the gatherer of data, Scott was happy to stick Chris Johnson in the role that he'd assigned to Stephen Kean—who was he to say one could run the machine better than the other? Numbers would speak for themselves. Besides, he felt prepared for the Lunenburg job. Not only was the building shorter than the one in Laramie, it was fully framed and structurally sound. He and Nate and Chris Raddell had carefully laid out a plan with Fernandes Masonry (which was struggling amid a shortage of masons) and gone over expectations for the speckled gray walls going up. Included in this outline was a plea for patience: With training and setting up and the learning curve, masons ought not expect big numbers right away. *Yeah, yeah*, the owners said.

Still, from the moment they parked the trailer on Monday morning, Zak and Chris Johnson hustled. It was 85 degrees by nine a.m. Shortly after, Scott rolled up in his own car and got to work, too. They unfurled one of the tents they'd brought to Vegas and set up a flat-screen in the shade. At ten-thirty, Scott led a short presentation. Julian Fernandes, one of the owners of Fernandes Masonry, was there, in a red Budweiser T-shirt, along with five of his crew. His guys grabbed a block each and, as soon as they took seats, started joking around.

"Is this the beach?"

"Is this a barbecue?"

"Where's the beer?"

"Take yooah fackin' foot offah theah!"

Scott barreled on through an exceedingly swift introduction to SAM while insisting that there was no rush. (Hearing this, one of the guys said, "Hey, I get paid by the hour, so you can take as long as you want.") He apologized for acronyms and pointed out parts of the machine: the generator, the propane tanks, the battery, the electrical cabinet, the bumpers and safety wheels. "Here's where you feed mortar," he said. "Here's where you feed bricks." He gesticulated wildly, sunglasses dangled from his neck, his yellow Under Armour T-shirt already drenched in sweat. "Every once in a while, this thing will chirp at you, and Chris will help with that . . . You'll get used to not blocking the laser . . . Be careful when loading mortar—don't put your feet under. There's no safety guard on this side . . . Any questions?"

Scott talked so fast it was hard to tell if he'd had too much coffee or just didn't want to restrain the crew. He certainly didn't want to bore them. "Once SAM is running, there's not a whole lot you gotta do," he said. "SAM can just crank." He paused a moment before adding: "You guys are responsible for quality. If things are off, you can adjust them. If a brick looks off, work with Chris."

By one p.m., SAM had been forked up onto the Hydro-Mobile, and Zak and Chris Johnson had begun mounting the smart and dumb poles on the north side of the building. By two p.m. the laborer had grown antsy. The crew's days ran promptly from seven a.m. to three p.m., and given the ninety-minute drive back to New Bedford amid Boston-area traffic, they did not dillydally when three p.m. rolled around. Finally, at two-thirty, Chris Johnson turned a key and fired SAM up so he could start taking measurements. He did so carefully, since windows and doors and corners "as built" were always slightly different than architectural drawings—and masonry was what hid the accumulated errors of all the other construction trades. This, of course, was one of the great services of brickwork.

Here, amid the two dozen measurements, some trouble emerged. Chris Johnson's wall map, based on the architect's plans, called for twelve-inch bricks and six-inch bricks, but the walls to SAM's left and right had not just twelves and sixes but tens, eights, and fours. Asked about this, the masons said they'd made the change because two-inch bricks, needed on the ends, looked funny. This annoyed Chris; SAM had only two chutes for two different sizes of cut bricks, and the change left him in need of a new wall map before he could slam any bricks in the wall. So much for planning. Long after the masonry crew had departed, Chris Johnson stayed up late, making a new map.

He began measuring for the second time at seven-fifteen the next morning and was immediately confounded by the lintels above the windows, which were not where the plans had them. Then the laborer admitted that he hadn't actually measured his cuts. Four inches was about four, six inches was about six. Chris Johnson grew more flustered than the day before. "This is all shit that was supposed to be figured out," he said. Like Fernandes's guys, he wanted to start laying. Instead, he threw down his tape measure, shook his head, and retreated to his laptop under the tent.

By the time Scott showed up with a box of doughnuts at a quarter after nine, SAM had just begun laying bricks—which made the chances of achieving twelve hundred that much less likely. Worse, the mud was soupy and not sticking well. By the third course, Chris Johnson worried about tipped bricks. Something was miscalibrated. Chris Johnson nodded, annoyed at SAM. Nevertheless, the brick count steadily rose.

On the fourth course, one mason said to another, "This might make my job go away!" Chris Johnson, having heard such prophecies in Las Vegas, said, "Nah." The mason, though, didn't believe Chris Johnson, and revealed a speedy wit. To the other mason, he said, "Hey, you know they were gonna call him SAM-J—Steal All the Masons' Jobs! Or they were gonna call him SAM-H—Send All the Masons Home!" Chris Johnson had no comment.

And then, from a modest construction site in the quaint woods of New England, a foreign sound rang out. One of the masons exclaimed, "*Jun caralho!*" Translated loosely from the Portuguese, this equated to "What the fuck!?" The mason scowled at the wall in front of him, as if it might respond with an apology. "Eh, what's this?" he said. "Stop the fucking joints, that's it . . . nah nah nah, it's fucking junk! I don't like that. Take it out. TAKE IT OUT!"

A crew of masons was working the old-fashioned way to SAM's left, and at first Chris Johnson couldn't tell if this mason was pissed off at the humans or the robot. Either way, the whole wall—human- and machine-laid—was one inch too high, so Chris Johnson agreed to take out ⅓₂ of an inch on the beds of the subsequent courses. Adjusting the tabs on both poles, he muttered, "My God, we might need to do drug tests at work. Those tabs were way off. Who did that?" SAM did only what it was told.

Meanwhile, the angry mason, whose skin was so sunburned it was the shade of baked clay, tore out bricks with a vengeance. "See this?" He grabbed a brick with a chipped corner. "Junk." He tossed it on the steel deck of the Hydro-Mobile. "This? Junk." He tossed it on the steel deck, where it made a clunk. He looked at the laborer. "Hey, you see this? Take this out. Take this out now." He had standards, that was for sure. He also had three decades of experience and more speed than anybody in Virginia or Laramie. He should have been in Vegas. As he laid and struck, he said to himself, "Rock and roll, rock and roll!" Thirteen seconds didn't impress him.

By ten-forty-five, SAM could reach no higher, so it was time to pause the machine and raise the Hydro-Mobile a few feet. Chris Johnson jogged SAM to the middle of the scaffold, then took a position on the right while Zak stood on the left. The Fernandes crew—who were familiar with Hydro-Mobiles—opened up two small trapdoors in the deck and fired up the scaffold's engines, which hid just below. These sounded just like lawn mowers. Then, in tandem, manipulating small levers, they began to drive the deck up the square towers. As the deck and everything and everyone on it ascended, the laborer, who should have been paying attention, failed to notice that the Hydro-Mobile had caught the bottoms of both poles. Amid the noise and motion, both poles were simultaneously wrenched from the wall. The right pole, held in place by an upper mount, stayed put, but the left pole—the dumb one—began to fall like a felled tree. Zak yelled, leapt to the side, and caught the dumb pole as it headed straight for the head of a mason. Thanks to his lacrosse skills, an injury (if not a lawsuit) was averted.

The incident left the masons aghast and bricklaying on hold while repairs could be made. On the ground, Scott wrenched on the bracket, trying to get it free of the expensive extruded pole, and it began to dawn on him that the day he'd had in mind was not at all what he was going to get. Meanwhile, up on the Hydro-Mobile, so much mortar spewed out of SAM's nozzle that it seemed less poop mode than diarrhea mode.

By the time Chris Johnson and Zak returned to SAM, the mud was hot and kicked, so they pumped it out into five-gallon buckets. At the same time, the speedy mason grew ever more eager to get going. "Hey-hey-let's-go-rock-and-roll," he said, as if waiting for the band to pick up their guitars. Chris Johnson's frustration level rose. He was annoyed that the tabs on the dumb pole were off, annoyed that Fernandes's crew didn't "know what the fuck they're doing," and vexed by the robot. Only a few bricks into the next course, Chris Johnson seemed about to explode. SAM was only half-buttering bricks. "Come on!" he said. "Why does it keep doing that?" That was stage four of debugging.

While Scott tried to help, an HVAC worker leaned over from a second-floor window opening and said to nobody in particular, "Hey, that thing's slow." Under his breath, Chris Johnson muttered, "Fuck you."

At noon, the fast mason saw something else that fell below his standard. "Look at this shit—can't leave this shit. We gotta tear it down. Turn it off. That's it—no more of this shit." He threw down his tape measure. "Don't play games with me," he said authoritatively. "We have to fix it now."

Chris Johnson tried to respond politely. "You want the robot to do it or you lay it?" The mason didn't answer. Instead, he tore out three bricks and told the laborer to shut up. Somehow, both foreman and superintendent noticed the situation and clambered onto the Hydro-Mobile to investigate. "If it doesn't look right," the superintendent said to the foreman, "it's on you." That was exactly what Scott had said. The mason clamped two bricks together, demonstrating for the foreman how thin the beds were. "We can't do this," he said. And then, to ensure that his foreman understood his agenda, he said, "You know rock and roll? Let's go! Seeeoooooww! Not like a turtle, you know?"

All of this Scott observed calmly while rolling a marble of mud between his right thumb and index finger. Chris Johnson's emotions were less subtle. He sat down on the track, removed his hard hat, measured the wall, shook his head, and put his hard hat back on.

The wall was still three quarters of an inch too high. Chris Johnson told the speedy mason as much, and the mason agreed. Then Scott stepped in. "You're right," he said to the mason. "Of course you're right, you're the mason. So, you wanna do it?" The mason said okay. SAM sat idle, its generator purring, its mixer mixing, and its arm not moving, while Scott showed the Fernandes crew, with a level and tape measure, how to adjust the tabs. At the same time, he steadied Chris Johnson's nerves by relieving him of one annoying task. To the masons, Scott said, "This is the most important part of the process—we want you to get comfortable."

When they finished adjusting the tabs, the fast mason flipped his trowel around his thumb like John Wayne did with a revolver. SAM resumed, and the men picked up with their striking. The fast mason lifted and reset every brick on the top course, tapping all with his trowel. He insisted the machine was too slow, but he was working hard to keep up. It was to little avail. By two-forty-five, the total brick count stood at 375. Not once did Chris Johnson mention his goal of twelve hundred bricks.

Ten minutes later, Julian Fernandes climbed onto the Hydro-Mobile and sat down on the track, facing the wall. He touched a few spots and said, "What the fuck? It's got way too many steps."

Zak tried to be considerate. "Too many steps? We could fix that with laser calibration. You gotta let us know."

The superintendent showed up. "How'd SAM do?"

Julian Fernandes answered: "Not good."

Team CR got a late start laying the next day because first they had to raise the poles. A new mason showed up, and he was as impatient as the John Wayne wannabe. "This is takin' forever," he said at nine-thirty. "I wanna lay bricks. Let's go let's go let's go!" At this, the laborer said he fuckin' hated the smaht pole. He wanted it to be *smahtah*.

Chris Johnson tried to do what Scott had done. "You guys wanna do this measurement real quick so we can get up and running?" he asked. But he didn't have the right touch. Hearing this, the laborer said, "Why do we gotta measure? Can't it just run?"

So Chris Johnson measured yet again and set the miserable tabs. Smudged with mortar, they were hard to move and nearly impossible to move precisely. Grudgingly, the laborer joined and grew so exasperated that he threw his cigarette to the ground. It was ten-forty-five, and the brick count was zero.

At eleven a.m.—halfway through the workday—Zak finally shoveled mud into SAM's hopper. Somehow, the poop bucket had vanished, so SAM just pooped mud onto the planks. With only four hours left in the day, the most they realistically could have laid was eight hundred bricks—because even though SAM ran at 275bph, the machine's long-term operating average was down between 150bph and 200bph. Eager to hit a respectable number, Chris Johnson skipped lunch and, as he manned SAM's controls, wished he could modify the tablet. SAM's software required its operator to log every bit of downtime so that the engineering team back in Victor could assess the issues interfering with SAM's productivity. The way Chris Johnson saw it, responding to the downtime logger was doubling his downtime, which pissed him off.

Two hours into laying, one of the masons said the bricks weren't coming out right. The other mason agreed. "If you gotta nudge something," Chris Johnson said, "go ahead." The mason did not want to nudge anything. He wanted his boss to inspect the wall.

The masons called the foreman over, and while he examined the wall, they yapped in more rapid-fire Portuguese. The foreman squatted, measured, hit the front of his hard hat with the tape measure, then checked the wall's plumbness. The foreman found a brick, tapped it, smiled, sat down, removed his glasses, looked at the wall, measured again, leaned forward, rested his head on the wall, and measured again. Then he made a few phone calls.

Chris Johnson, meanwhile, rolled SAM to the middle of the Hydro-Mobile and stretched, making clear that this time, he was the one waiting. Momentarily free, he found and reattached SAM's poop bucket.

An hour went by. The laborer cracked jokes. Zak sat and endured them. With little else to do, Chris Johnson emptied the poop bucket, whose contents rose so high they suggested a spire by Gaudí. SAM's count for the day was 294, and it would rise no further.

Julian Fernandes delivered the ultimatum at eight the next morning. The machine was slowing the job down, losing him money. He wanted SAM off the job.

Zak and Chris Johnson called Scott and Nate to deliver the news. "I feel like I'm chasing my own tail," Chris Johnson said. "If they don't tell me when things are wrong, I don't know." Zak thought one of the masons was deflecting blame on SAM to cover his own ass. Scott's response was disbelief. Certain he could turn the situation around, he instructed Zak and Chris Johnson to develop a "game plan" for the project manager. He insisted that the mason needed to get comfortable working around SAM, and said they needed to "see what needs fixing—either the machine or the mason."

Beyond the walls of the trailer, it was a beautiful, clear day—the nicest of the week, and for the nth time, bricks were not going in the wall.

Fernandes's project manager showed up at ten-thirty and said he wanted to see SAM run. With that, even though Julian Fernandes had said "no more," the men recolonized the Hydro-Mobile. As they ascended, Zak told Chris Johnson he'd better be on his game.

Chris Johnson shoveled mud into SAM's hopper. Zak loaded bricks onto SAM's conveyer, then cleaned out SAM's nozzle. The foreman barked in Portuguese for the other guys to get over there. SAM *chirpchirpchirped*, alerting men not to block the laser.

The project manager squatted. He held the temple of his glasses, inspected the wall, checked it with a level. Meanwhile, the fast mason found a banana brick on the top course and tore it out. Fernandes's crew, in charge of quality, should have caught it the day before.

The project manager took issue with the mortar. He said it wasn't grabbing well. Chris Johnson could have said, *You're right, and we should talk to the man mixing your mortar so we all end up happy*—but he did not. Instead, he simply said, "I'm workin' on it."

Under scrutiny, SAM laid two courses of bricks. Now the project manager said every brick was sawtoothed and did not like that his masons had to touch every one. Again, Chris Johnson said he was trying to fix it.

The machine pooped. Its mixer squeaked. It proceeded to the third course. The fast mason smiled. And then, just after noon, one of Fernandes's other owners made his decision clear. "Forget about it," he said. "It's not gonna work."

So halfway through their third working day, they packed up. As Chris Johnson and Zak removed the poles, the masons continued building the short gray wall, laughing and telling stories as they had been all day. One of them stopped Chris Johnson and asked him to move the machine out of the way. They did not say goodbye, or give any condolence, or offer to help carry equipment off the scaffold. Chris Johnson thought: *This industry will not be modernized.*

What followed in Victor was CR's second great come-to-Jesus reckoning. As before, the engineers unanimously doubted SAM's viability. To Chris Johnson, mimicking people seemed an absurd proposition, and he questioned the validity of the whole endeavor. Tim told his boss that using SAM came with too many situational caveats. Just setting up—installing the track, mounting the poles, measuring the wall—seemed too difficult to him. "I wouldn't buy it," he said. John figured the wonky Wi-Fi was the machine's greatest flaw and thought letting masons run the machine was untenable. Kerry told Scott that SAM wasn't user-friendly. Mike Oklevitch said SAM wasn't intuitive. And Stephen Kean, whose sole responsibility was running SAM on jobs, and who had coaxed the machine to its highest-recorded performance, called it unreliable. Even Scott, for his part, knew SAM still looked—and behaved—like a prototype.

And so two Big Decisions were made. First, recognizing that their machine—whether or not it was a lemon—was subject to a learning curve at least as steep as that of a stick shift, they decided that training was key. A teenager didn't just get in a car and drive to Tijuana, after all. He had to learn how to drive. Henceforth, masons would learn, there in Victor, how to work with and around SAM. Second, though Scott didn't put it in these words, he could see that SAM's productivity didn't justify all the trouble that working with the machine entailed. For whatever reason—because it was a robot?—SAM was held to a higher standard. And SAM didn't yet meet Scott's own standard. He still dreamed of production at 360bph, while SAM had barely hit half of that. So he pushed, as he had been pushing for years, for a faster machine. To motivate his engineers, by way of his best swim-meet pep talk, he said that it was darkest before dawn.

He was suggesting that the light of day was close.

Pressure was running higher than normal, because a bricklaying robot in development halfway around the world had just made its debut, and now everybody wanted to know which robot was superior: SAM or Hadrian. Thanks to a patent, Scott had been aware of the Australian machine since 2009, growing increasingly anxious about it. The Aussies behind it, who had aerospace work in their past, had kept hinting that they were building a machine that could lay one thousand bricks per hour. *A thousand bricks per hour!* That was one every few seconds—so fast it would lay *half a dozen* bricks to each of SAM's. Photos had shown an elongated front-end loader with a conveyer belt running up its arm and a gripper in place of a bucket. In its four paws, it was grabbing a red brick. Where SAM looked like a hot dog cart, Hadrian looked like a veritable Brachiosaurus. Then Fastbrick Robotics released a rendered animation, purportedly showing Hadrian building an entire house in two days. The media, of course, leapt all over it—and as a result, Fastbrick soon rounded up some serious funding.

The video was a great relief to Scott, because until he saw it, he couldn't tell what the Aussies had actually built, or if their machine had been on a real jobsite (which he knew was the truest test of all). Finally, he had an answer: There was no mortar, and there were no bricks. Instead, Hadrian assembled blocks, using a thin, modern adhesive. Each block, Fastbrick figured, represented eight bricks—hence the amped-up, disingenuous "brick-equivalent" speed of a thousand per hour. As such, Hadrian—more concept than product—was no more impressive to Scott than ROB, and years behind SAM. Scott just hoped the construction industry would recognize as much, which was hoping for a lot, considering the miserable start SAM had gotten in its first construction season.

That summer, as public interest in SAM mounted (few media outlets could resist a wacky robot story), the engineers homed in on SAM's motions the way a good coach attended to an experienced swimmer. When exactly was he rotating his head to breathe? How far was he rotating? Where were his hands exiting the water? Was he flicking his wrist? With an eye toward refinement, they studied how SAM held and placed bricks: how hard it had to squeeze so as not to lose its grip; when exactly to smoosh mortar onto each brick; how long, once it had placed a brick, to vibrate the gripper and hence cause the mortar to settle; and how to calibrate the whole sequence. They reasoned that to get the most out of SAM, the Stäubli could reach high enough to lay eight courses before the deck of the Hydro-Mobile had to be raised a few feet—but that it was faster if SAM reached only to the seventh. And while Scott inspected practice walls at the office with a ruler, he took his studies further after work. He took videos of SAM's performances and examined them in slow motion, like a modern umpire. Often he did this at home, before his dining room table: He watched slow-motion videos of a machine buttering a brick.

Chris Johnson, meanwhile, was working to eliminate redundant micro-motions from the Stäubli and to whittle down SAM's cycle time. Every time Chris Johnson took a hair off the cycle time, John Nolan had to make sure the command loop synced with the rest of the machine. Every time Mike Oklevitch tweaked the mortaring, John Nolan had to check that all of SAM's neurons fired properly. John Nolan at least was getting very skilled at the continuous iteration demanded by his boss's approach.

At some point, all of that abstraction apparently got to John, because one afternoon he decided to actually build something with bricks. He'd tooled a lot of SAM walls but never laid bricks of his own. After lunch, he headed out back, toward the highway, and built a barbecue grill. Mike Oklevitch mixed up a batch of real mortar for him, and Glenn came to help lay bricks. They used big red Beldens, about three dozen in all—and together, the engineers averaged ten bricks per hour. No foreman complained about their quality, and Scott didn't complain about their pace. They started using the grill as soon as the mud hardened.

16.

Washington

Aside from having to spend August in Washington, D.C., the month went well for the engineers: On a handsome three-story school there, SAM finally broke through its glass ceiling of 150bph and ran at 250bph for five straight hours. It helped that Clark was the general contractor of the job, and that the masonry contractor was the same firm that had so eagerly thrown SAM at the army barracks nine months earlier. Together, they cut out the shenanigans that had interrupted work in Laramie, and eliminated a lot of the farting around that seemed to be par for the course in construction. SAM was their priority, and they made sure not to hold up the machine. Masons kept a stack of cut bricks on hand and ensured that fresh tubs of mortar were always available. Clark also made certain that masons—the *same* masons—always showed up at work to take positions beside SAM.

Aside from two small sections of gray bricks hand-laid in the stack bond (one directly atop another), SAM had the whole two-hundred-foot-wide north facade of the Lab School to itself—and this, too, made a difference. This facade was broken up into two parts: a sixty-foot section on the left and a 140-foot section on the right. On the smaller section, the architect called not for four-pound modular bricks but for six-pound economy bricks, in "flashed red wine," laid in the running bond. On the bigger section, the architect called for longer, more slender (and just as heavy) Norman bricks, in "matt red," laid in the one-third bond. In front of each section was a Hydro-Mobile of equivalent length. Thanks to a new part of interlaced steel fingers that Tim had devised, SAM could smoothly traverse the second, super-wide, jointed Hydro-Mobile, giving the machine the freedom to attack huge high-potential walls. It was this second wall that got Scott especially excited.

Before SAM attacked the big wall, CR had it attack the small wall—and it did so with a vengeance. On just its second day, SAM placed 1,364 bricks and, soon afterward, placed 1,383. (Scott doled out no bonuses for hitting the 1,200-brick mark; instead, his threshold inched higher.) SAM, running on its fifth software version, seemed like a real hurrier: The Stäubli moved to the wall faster than ever, the gripper was nimbler, and the machine's operator could keep tabs on mortar consistency from the tablet rather than relying on a mason to tell him that the mud was no good.

Then SAM moved to the double-wide scaffold. Up on that giant steel deck, it was blazing hot—so hot that ice cream sandwiches brought by a sympathetic reporter melted in seconds. One afternoon, the laborer sat down on a stack of bricks, brought a water bottle up to his face to drink, and was so exhausted he missed his mouth. To Stephen, who'd put a fan on top of SAM, the work was way harder than anything he'd done in an auto garage. But SAM didn't mind the conditions. At the top of the second floor one day, the machine laid 1,486 bricks. As it became evident that the day's count was going to be huge, Clark's superintendent climbed into a balcony and called the brick count "friggin' awesome." Stephen Kean was elated. Chris Johnson beamed. So did Scott.

By the end of the D.C. job, SAM had placed eleven thousand bricks in eleven days, and all signs pointed to the conclusion that it could have done so faster; at least three times, the brick count had fizzled on account of some snafu. One morning, the engineers couldn't find the key to turn SAM on bright and early. Another day, a loose wire cost SAM two hours of run time. Once, the generator stalled and caused an electrical brownout that put an end to a good laying day. Such things grated on Scott's nerves.

But by and large, CR could declare a victory. Mike Oklevitch had befriended the mortar mixer, which averted any sudden flare-ups over the quality of the mud. The other engineers had always stayed as late as required to set up the poles at the end of the day. And when a new wall map was required, they'd been able to retreat not to an industrial trailer but to an air-conditioned office inside the building. They kept their cool, and the result was SAM's sharpest building yet. The head joints, all ramrod-straight, did not look strange, as architects had once predicted they would. They looked dashing.

In fact, it was almost too bad that SAM had wrapped up before the school year started, because if anyone was primed to appreciate the revolutionary machine built in a revolutionary manner, it was the students and teachers of Washington's innovative, unconventional Lab School. Still, it was hard to complain when appreciation for quality had been supplanted by appreciation for speed. The foreman, who had never placed more than eight hundred bricks in a day, was impressed by SAM. The lead mason—who'd never laid more than twelve hundred—was, too, even if he phrased it strangely to Zak (who was trying to gather marketable blurbs): He said SAM wasn't threatening his job so much as making him work harder.

A month later, in Maryland, Scott earned two more business victories.

First, SAM attacked the wide wall of a brick distribution warehouse and placed four thousand bricks in a twenty-four-hour run. The job went so swimmingly that it almost made up for the *disastrophes* in Wyoming and Massachusetts: Not only had bricklaying through the cool night and gentle fog seemed downright peaceful, but the economics were self-evident. Where the project under typical methods would have taken four days and cost $1.50 per brick, it had cost only a third of that. Better still, SAM installed a spiffy two-tone logo in the wall, a big PVB, for Potomac Valley Brick. PVB was the biggest brick distributor in the D.C. area. PVB's owner, who had chaired the Brick Industry Association, knew good brickwork when he saw it, and was connected to architects and developers. All of this boded well for Scott. But what really excited him was that SAM ran fastest—at 240bph—with just one operator and one mason.

The second victory involved a breakthrough with the union. The president of the BAC, Jim Boland, having visited Victor personally to check out SAM earlier that summer, had invited Scott to the 150th annual convention of the International Union of Bricklayers and Allied Craftworkers. This was in Baltimore, at a packed convention center. At first Scott had found himself in the lion's den of resistance. Sensing a threat, one mason said to him, "I've been laying bricks for thirty-five years, and if a robot told me where to lay bricks, I think I'd shove it off the scaffold!" Scott quickly clarified the matter of who was in charge, but appeasement remained nil. A Bostonian said he hoped never to see SAM in his city. Someone else said, "This is the first robotic bricklaying system I've ever seen, and I hope it's the last." To Scott's face, others called SAM scary as hell, valueless, and far from ready, and said, "I hope I'm retired before I see this on my jobsite." They said it wasn't worth using and that it'd never work.

At a demo in a parking lot, where Scott tried to show union guys that it did work, they only busted his balls more. *Your quality looks like shit*, some said—which hurt, because the machine's quality was better than ever. *You're stealing jobs*, said others. (Scott often countered that SAM would bring about increased masonry work.) *This technology is just a fad, and it's making everybody soft*, they said. "Where's your union card?" one man demanded. It was like that very first job all over again.

Even when the union president, in the main hall of the convention center, showed a video featuring SAM, there were grumbles in the audience. It didn't help that the president of the AFL-CIO had just gotten the crowd riled up by thundering on about an economy that was rigged against workers. But then the president of the bricklayers' union stated clearly that the organization needed to adapt if it was to survive another 150 years. Union bricklayers, he explained, had to learn from the past without living there. So, he said, they had to embrace SAM, claim running it as a job for bricklayers. The union, after all, had been looking at bricklaying machines for a century—and they weren't mere notions anymore. He put it this way: If he were a kid, he'd want to work in a trade that embraced new technology, and if he were an older bricklayer, he'd rather have a robot helping him than not. Scott was blown away by the magnitude of the statement. He called it a huge step, and notwithstanding the jabs from so many masons, he no longer felt entirely like a black sheep.

Soon afterward, Jim Boland himself told Scott that the union officially intended to embrace SAM, and invited him to a Florida gathering with the executive committee and a hundred of the union's signatory contractors, some of the country's largest.*

Leaving Baltimore, Scott was thrilled, because even if union bricklayers didn't have his back, union leadership did. He and Nate could have danced a jig. Nate later called James Boland progressive. Attitudes of bricklayers, both prophesied, would come around. It was inevitable. The rest, they figured, was details.

Back in the office, though, Scott felt disappointed with his engineers. They did not yet have the setup process dialed—and because the hours spent setting up were hours SAM wasn't getting bricks in the wall, and hours during which bad impressions formed, Scott wanted improvement. "We gotta fix this," he said. He called measuring "death by a thousand cuts," and the transition from poles on the wall to bricks in the wall "a black hole that is death to us." The stoppages represented potential bricks vanishing into the ether.

All his engineers could come up with was shorter poles, which would make setting up a one-person job but necessitate more setups; or mounting the pole brackets closer to each other, which could result in wobbly poles; or using a string line, a suggestion so regressive that it irritated Scott on principle.

All of their brainstorming, Scott thought, was just blah. What he wanted was to set up and tear down the poles a hundred times and optimize the procedure. Better yet, CR should make a mechanized bracket that used software to position itself. Why hadn't his engineers thought of that?

The engineers did not embrace this idea. Such a bracket might work, but the real solution, they knew, was the quarter-million-dollar laser tracker made by Leica. With it, they wouldn't need the poles or the brackets or the finicky laser box at all. All they'd need was a bunch of money.

*Scott never told the union president that his kids had the initials B.A.C.—in that order.

17.

Home

And then Scott was working in his own backyard with a masonry firm called Brawdy Construction. When Chris Raddell had first called Brawdy to ask if they'd heard of SAM, Brawdy's number two guy, a young estimator named Ryan Glenn, had surprised him by saying yes. Ryan had first heard about SAM when the machine was declared the most innovative product at World of Concrete. He'd watched videos and mentioned SAM to his boss, Jim Brawdy, and though both were intrigued, they had doubts. Soon after, Ryan and Jim sat down with Scott and Chris Raddell and found they couldn't trip the new guys up. CR had thought of everything. Ryan was impressed.

Scott, for his part, couldn't believe his luck. Over a year, he'd been hunting for what Nate called "astute" and "visionary" masonry firms to work with—and here was one just down the street from where he'd grown up, veritably hiding in plain sight. In an otherwise union town, Brawdy was a non-union firm. Brawdy, the biggest masonry firm in western New York, had a hundred employees, mostly young, capable guys who weren't intimidated by new technology. The company did about $8 million of work annually, making them big enough to recognize the potential in SAM. Brawdy had built Buffalo's Barnes & Noble and Citibank. At the end of September, Brawdy rented SAM for two thousand dollars a day to put in sixteen thousand bricks on the new headquarters for Columbus McKinnon, which had been making hoists and cranes since George Eastman had begun making film. It was a two-story building in the shape of an X, and Brawdy had until the end of November to get the walls up. At the start, Ryan Glenn uttered eight words that made Scott's heart sing: "We're hoping we get through the learning curve." This gave Scott hope that small, patient, progressive masonry firms really did exist, and that if there was one in Buffalo, there was one in every city.

Having put Stephen Kean at the helm in Laramie, and Chris Johnson at the helm in Lunenburg, and then thrown nearly the whole company at the jobs in D.C. and Maryland, Scott was back to relying on Stephen. With help, Stephen was set up and running by the end of September. He got four days in before it rained so much that the region was under a flood watch (and the leaks in the office roof worsened). The mixer was behaving oddly, and the Wi-Fi connection between the laser box and the machine was cutting out, but otherwise SAM ran well. During those first days, the machine averaged 550 bricks daily, which was better than it sounded, because it was putting in big ten-pound utility bricks. On the other hand, SAM never ran longer than a few hours on any of those days, and this disappointed Scott because he knew SAM could do better. He called a meeting.

As Scott put it, the wheels kept falling off; instead of getting six hours of run time, Brawdy's crew was barely getting half that. As a result, SAM couldn't hit the thousand-brick mark. Scott wanted all hands on deck, by which he meant all engineers at the whiteboard.

John, Glenn, and Tim made the case that CR needed to stop looking so hard at "bricks in the wall"—not just because CR was doing so much other work (setting up Hydro-Mobiles, anchoring Hydro-Mobiles, installing brick ties), but because while six hundred bricks might look great one day, twelve hundred might look great the next. Windows, doors, lintels, bands of stone: All could interfere with production, and deserved mention. To Tim and Chris Raddell, Stephen was working so hard that it looked like Construction Robotics was in "survival mode." John explained why the wheels looked to be falling off: On any given day, Brawdy's crew ran out of cut bricks, and because cutting bricks was a task that fell on the foreman, who was otherwise busy, the mason had to scamper down to the ground to cut more. In the meantime, SAM kept running, so by the time the mason returned, he was way behind striking joints, which inevitably resulted in Stephen letting SAM sit idle while the mason caught up. To John, this choreography was difficult and unsustainable. Tim backed him up and astutely suggested that it would be nearly impossible to convince a masonry firm that its own crew wasn't operating properly. The onus would always fall on SAM.

Scott figured smaller steps could solve the problem. He declared his wish for Stephen and the foreman to come up with a "clear plan" for every working day, a list of exactly how many cut bricks Stephen needed up front, how many tubs of mortar he expected to need, how many basket lifts he'd need access to and when. Henceforth, Scott wanted Brawdy's crew to clamber into said basket lifts and raise the poles and take measurements every afternoon, so that SAM and the masons could be up and running at 7:05 the next morning. He called such an arrangement "a perfect day," and had faith that Brawdy could get in the routine.

But routines were just that, and changing them took effort and time. At the end of the workdays, the masons were in the habit of cleaning up, not of doing more work and thinking ahead. At the end of the day, they were winding down, done thinking. The construction industry, the engineers began to learn, was as slow to change as baseball. Masons didn't come up with excuses, and they didn't have much sympathy for engineers who did. So: some innovation!

Regardless, two days later, other matters took precedence. In cold severe enough to warrant leather gloves and wool hats under hard hats, SAM's mixer stopped working. The gearbox was stripped. When Tim, having wrenched on the thing for thirty minutes, called Scott to discuss the logistics of a repair, neither one quite said it: Sure, the broken part was confounding, but at least it had happened on a Friday afternoon. CR could fix it over the weekend and settle into a new rhythm the following week.

The fix fell upon Tim. Late Saturday night, having somehow processed the booze of a bachelor party, he rebuilt the gearbox. There was a benefit to working close to home, and to having no security on-site.

The next week, issues manifested in the Stäubli arm. Usually, when the Stäubli was confused, it would stop mid-placement, open its gripper, and drop the buttered brick as if it were a dirty diaper. Such bricks thunked onto the planks, splattering mud in a small blast zone. Sometimes the Stäubli just froze and didn't open its clutches. Then the machine wouldn't respond at all, even after rebooting. With two courses down, Stephen got so many robot errors that he summoned Chris Johnson to Buffalo. Chris Johnson, the debugging maestro, found robot faults he'd never seen before, and pumped out SAM's mud before it hardened—putting an end to their day. The only solution was to swap the machine in Buffalo with the test machine in Victor, exactly what the MBA students had once suggested.

Through rain, Zak drove the shop machine west. Tim got behind the controls of the telehandler and forked the broken machine from the Hydro-Mobile. Stephen, who filled the time by looking at yet another jalopy, returned to the site and double-checked that the replacement machine worked fine, then discovered that it wouldn't connect to the Wi-Fi network. He called Chris Johnson, who called John Nolan, who was in bed. Eventually, they figured out the problem and completed the robot swap. The engineers returned to Rochester after midnight.

The following day was SAM's best day yet. Stephen squeezed four hours of run time out of the day and got 936 bricks.

In the shop, things went less well. The Stäubli error magically disappeared, so when the tech arrived to see what was up, there was nothing to troubleshoot. Scott, about to head to Florida for a vacation, remained unhappy. It did not seem like a good time for a vacation—not with mysterious problems and an unfinished job on his mind—but it never seemed like the time was right. The only good news was that there was some swimming in his future.

The robot error returned a couple days later. Nate swung by the office and told the engineers to do whatever was necessary to avoid stoppages. "Anything you need," he said, "do it." (What they needed was a Leica, but they didn't say it.) SAM's production numbers remained between five hundred and seven hundred, even as Brawdy's masons staggered their lunches so that SAM could keep laying. Tim had an unpopular but prudent idea: to replace the bearings on the pump input shaft before they failed.

The temperatures dropped below freezing, making the good Stäubli act sluggish. Something was wrong with Brawdy's Hydro-Mobile, too: It was twisted, leaning so much on the left that it scraped up against the wall. As a result, Stephen limped along on the right half of the scaffold until Brawdy ordered a replacement.

Hoping for a nudge in the right direction, Stephen pushed updated software into SAM, because the old version wasn't buttering half-bricks. The new version, though, buttered the corners between head and bed joints poorly. So even on flat, windowless sections, Stephen failed to climb much above seven hundred bricks a day. It was painful—more so when Brawdy's telehandler operator vanished, leaving SAM mudless, or when Stephen ran out of cut bricks again, or when the laborer failed to show up for work. The wheels that had seemed so round in Washington, D.C., kept falling off.

Over the weekend, snowflurries fell.

During the last week of October, Scott—now back in New York—clung to the thousand-brick dream, hoping to impress Ryan Glenn, even though Ryan insisted he was okay with the current numbers. It was a tough job, and the pace was not disappointing him. But the numbers frustrated Scott. For him, watching the job proceed was brutal, because the crew regularly hit five hundred bricks by lunchtime, then stopped to set a band of stone and ended up putzing around so much that they never picked up their earlier rhythm.

On Stephen's birthday, he stayed late to top out a wall section, enduring a choppy Wi-Fi signal. The engineers tried a new Wi-Fi box on SAM, but it made no difference.

At the office, the engineers scheduled a meeting with a Wi-Fi expert, but his tweaks didn't help, either. They needed a spectral analyzer. Chris Johnson remained unable to locate the robot issue. It rained like hell. Then it cleared up, just in time for a site visit from a mason in Pennsylvania. On the last day of the month, SAM got in 860 bricks, greatly impressing the visitor. Stephen, though, was less than thrilled to see variation in the head and bed joints, and he summoned Glenn to Buffalo. The wall was half an inch higher than it should have been. Something seemed miscalibrated.

The job stretched on for two more weeks. During this time, Scott started listening (he only did audiobooks) to *The Wright Brothers* by David McCullough. The incremental development of the airplane reminded Scott immediately of SAM: He knew that was how invention really worked. "There's a little bit of soaring—ten seconds, then more, then more, then improved flying, then an engine on it, and years pass between the first time they're airborne and when they're actually selling the idea," Scott explained. The press wouldn't pay attention because flying machines seemed like colossal failures, and who were two Ohio dudes to claim otherwise? Yet to the two men, they'd just invented the greatest thing in history. Scott couldn't get over the parallels.

During this time, Stephen developed as much doubt about SAM as he'd developed about the corona. As he grew less enamored with SAM, he grew more excited about an automatic brick-cutting saw, a product that had more obvious appeal. He spent a day at the office, detailing all the controls that needed fixing so that SAM could be handed off to a customer. He wrestled with Wi-Fi so spotty that SAM buttered a brick, reached out, and then waited thirty seconds before reconnecting. He endured the failure of the bearing that Tim had suggested fixing weeks before. He endured heavy rain and Laramie-like wind. But on the twenty-ninth day, he wrapped up the job.

When Scott drove out and analyzed the job with Jim Brawdy and Ryan Glenn, both insisted they were pleased—even though, during twenty-nine workdays with SAM, only twelve days involved three hours of production, and only two barely reached four hours. Average production time on the job was two and a half hours a day, for just above five hundred bricks a day. But Jim said that SAM had saved him ten thousand dollars on the job and that he'd love to keep working with it.

However, he couldn't justify buying the machine. Productivity just wasn't high enough, and Brawdy didn't have enough work to keep the machine busy. Jim Brawdy said he'd need four big jobs a year to warrant a purchase, and big jobs in Buffalo—hospitals, schools—tended to go to union firms. Jim and Ryan had the desire but not the justification. There was too much risk. If their company were twice its size, they said, they'd absolutely buy SAM. For now, they'd rent it.

Since Laramie, SAM had gotten 20 percent faster, but the machine still wasn't fast enough for Scott. Given the construction industry's unwillingness to embrace "the manufacturing mentality"—which would earn SAM seven hours of run time in a day—every hour was that much more precious and that much more critical. So within weeks of the Brawdy job, Scott called another meeting. The main topic was speed.

"We need higher productivity," Scott said to Zak, Chris Raddell, John, Tim, Stephen, and a new director of engineering. Two hundred bricks per hour was not enough.

That, Scott said, "looks like shit." He'd already run the numbers. Though a year contained 365 days, only 250 were workdays, and if you subtracted days consumed by setup and moving and rain, that left only 150 days in which SAM could crank out bricks. For the machine to pay for itself in two years, it had to lay 1,400 utility bricks a day and more than 2,000 modular bricks in a day, which meant the machine's cycle time had to be faster.

Chris Raddell agreed. Because SAM rarely hit the number he promised, the machine looked half built, half ready. (There was, of course, a term for this: MVP. And a solution: Don't overpromise.)

For CR to survive, Scott said, SAM needed to run faster. He wanted speed benchmarks. He wanted to know what to expect.

"When did thirteen point two seconds become not good?" John asked. "I agree we can get better, but we're approaching the limit of the fundamental architecture of the machine."

"What's our limit?" Scott asked.

"Eleven point seven seconds: three hundred bricks per hour."

"That's it?" Scott asked. "Where's that from?"

"From Stäubli! That's this motion at one hundred percent! We can get better, but the whole premise of this was two-twenty, and we've blown past that. So I struggle—"

Here Chris Johnson chimed in. "We hit eleven point eight," he said.

At this, Scott perked up. "If we hit three hundred bricks per hour," he said, "we're over two thousand bricks per day." He made it sound like they were so close.

John said that he and Chris Johnson could squeeze half a second more by optimizing the robot's motion, putting the machine at 11.5 seconds or 313 bricks per hour. "Other than that," he said, "you're talking major changes."

"That's huge!" Scott said. "You'd get over two thousand bricks per day!" If SAM really could lay down three hundred bricks per hour, he was sure that masonry contractors would figure out how to "squeeze the shit out of" SAM. The harder they squeezed, the bigger the numbers would get, and the lower the payback time would get, until SAM was undeniably dominant. He told his engineers to set targets for every size of brick.

Nate, cutting in by phone, had his own perspective on speed. He thought the interaction between CR's engineers and a masonry firm's crew slowed things down. Each, after all, had its own routines. The fastest way to run SAM, he thought, might be not with an operator like Stephen but with three well-trained masons.

Zak didn't say it, but he also thought running without Stephen was wise. He thought Stephen could have done more to familiarize Brawdy's crew with SAM—pushed them more and fiddled with the machine less. "He wants a perfect wall," Zak said later, "but there's no such thing as a perfect wall. There's a good enough wall."

On behalf of the overburdened engineering team, John tried to object but was drowned out by enthusiasm.

"It's gonna take off in the construction industry," Nate said. "This year, next year, soon. It's going to change. Right now, we're dealing with an archaic system. It's going to change."

"If we get two hundred and fifty bricks per hour," Scott said.

"Two hundred and fifty bricks per hour," Nate echoed. "It's all I give a shit about. Forgive my French. I don't care about job time. Just machine time. We're close."

"We're close," Scott echoed his father-in-law.

18.

Tennessee

With WASCO, Scott was counting on the weather. The masonry company was short on masons, having lost a legal battle with the union, and wanted fifteen thousand bricks installed on a school's facade in ten days. If the weather in Nashville cooperated, Scott figured he could really impress the firm and hit two thousand bricks a day.

The job, fifteen miles north of the city, was the first for John Lombardi, the newly hired field-services manager for Construction Robotics. The position had been created to liberate Scott and his overburdened engineers (who were building more SAMs) so that they might have time to consider the torrent of feedback pouring in from trainings. It was also the first job for a new young operator named Ken Carr. Like Stephen, Ken was a car mechanic, and like Glenn, he'd built a robot of his own design in school. Construction, though, was new to him. As such, Stephen was with them for what the army liked to call on-the-job training. The only difference was that the cadet was providing OJT for his colleague as well as his boss.

SAM started with a bang, placing over seventeen hundred dull gray bricks on the first real day. The machine had a new antenna, a new mixer motor and cooling fan, a new gripper, and new mounts for improved propane tanks. Its software was cleaner than ever, displaying over- and undersize joints in 3-D wall maps, usable by engineers and non-engineers alike. The software also streamed the machine's brick tally to a website, so that a certain speed-obsessed engineer could track the machine in real time from his phone. Even the smart pole had been refined: shortened from twelve feet to eight and reconfigured to use less battery power. SAM's cycle time was just over eleven seconds. Better still, SAM now did a little dance at the start of every course, to prevent nudging the first brick out of place with the second. Glenn and Chris Johnson had devised this maneuver and called it the "shell game." Instead of laying 1 2 3 4 5 6, SAM laid 3 1 2 4 5 6. By using the shell game, SAM lost a bit of time rolling from 2 to 3 and then from 3 to 4, but its bricks stayed where they'd been placed, and the machine looked downright smart.

On that first day, the mud—technically block mud—was awfully thick. Stephen called it sucky but made do. Then a historic weather front moved in. For two days, it didn't get above freezing. It was ten degrees colder in Nashville than it was in Rochester. Snow fell on the Grand Ole Opry. Late the next morning, the temperature finally crept up, and the antsy crew returned to work. All day, Ken was nearly cold enough to lay brick with his tongue—and the mud was cold, too. The mud was so cold it never became mud; the chemical reaction never began. The stuff, just sand and water, didn't cohere. A lot of cursing followed. Stephen called the scene an "absolute shit-fest."

Because fewer than average bricks went down, phone calls ensued. "There is a difference," John Lombardi reminded Scott, "between having a concept that doesn't work and having conditions that aren't right." So desperate was Scott to show his machine working that he considered the crew laying a long bed of mud and SAM placing dry bricks on top, but the February weather didn't allow mortar to be placed by machine or man. Anxious to get a good solid day of running in, Scott sent mortar master Mike Oklevitch south to get a handle on the situation.

With Mortar Mike, team CR rallied, hitting their mark, even if many of the head joints failed to stick. But Stephen and Mike Oklevitch found themselves engaged in damage control. WASCO's president kept referring to SAM as "in development phase"—which it was—but the phrase insinuated that it was barely a prototype. Scott, whose travel schedule was starting to look like that of the Destroyer, made plans to head south.

The next week began with a day lost to rain, but at least it wasn't snow. The next day ended before lunch because of more precipitation. By then, SAM was eight hundred bricks behind schedule.

As soon as Scott deplaned in Nashville, he texted Stephen: *Am I going into a shitstorm?* Stephen said not at all. SAM already had one thousand bricks down. Scott showed up on-site midday, donned a hard hat, and climbed onto the Hydro-Mobile, whereupon he found a scene so calm the masons looked bored. Mike Oklevitch had figured out the necessary mortar parameters, and even if some of the head joints kept falling off, adhesion was phenomenally improved. Plus, Stephen had found his rhythm. Rather than stop SAM to raise the Hydro-Mobile, he executed what he called a continuous raise, which elicited a funny look from Scott. The look said: *Don't mess this up now!* "Don't worry," Stephen said, "I know what I'm doing." And he did. SAM laid 2,261 bricks that day, a new daily record.

That night, Scott and WASCO's chief met at a steak house to discuss the job. Even if the masons were indifferent to the robot, and the project manager was unconvinced of SAM's merit, SAM had caught up.

Scott decided to push new software (CR's eighth version) onto the machine the next day, and it was a mistake. The process took forever, and SAM fell behind pace. Scott called it a black eye for CR and took a lesson from it. Henceforth, his engineers would push software only during evenings or weekends or in emergencies. Then the mixer ran away, which made Stephen want to revert to the old software version, which at least was glitch-free. Over the phone, he brought up the glitch with John Nolan. Give it a day, John Nolan advised. It was almost as if SAM had an immune system.

Quality fell off, too—not on account of the software but because the laser was bouncing. The smart pole, mounted to the roof, shook when roofers scuttled about. "Kinda eh" was how Stephen described the wall taking shape. Sections were definitely sawtoothed. Adding insult to injury, on a joint between two sections of an extended Hydro-Mobile, SAM got temporarily beached.

At least the weather was warming up. Then it warmed up so much that the winds began to howl, making the laser really bounce around. The mixer bug, self-healed, did not present itself, but something was off, because a control joint on the wall ended up out of plumb. It wandered like a vine. But the masons also flubbed the bond pattern on the edge of the wall, and hid their mistake behind a downspout, so they couldn't point fingers at SAM.

To make up for the weather, the masonry crew put in time on Saturday. On account of the thick mud, or maybe just accumulated wear and tear, the augur bearing failed and the mixer motor overheated before they made much progress. So the crew returned to the site on Sunday. Stephen replaced the bearing and improvised a fix for the mixer. Under SAM's brain, he zip-tied a hair dryer. He set it to cold, ran the stream of air through an old piece of plastic hose, and pointed it at the mixer. SAM was starting to look like the old Alpha machine.

Two days remained before Stephen expected to reach the top of the wall. On the penultimate day, the crew ran out of mortar mix, so they shook out the last bit from the silo and let it mix slowly for an hour. The result was what Stephen called the best mortar of his life. Euphoria, he called it. Breathtaking. It was veritable glue. Installing lintels slowed things down, but SAM held the pace. And then on the last day, SAM set another record, topping out the wall with 2,294 bricks. It was exactly the finish Scott had always wanted.

WASCO's chief, though, remained of mixed mind. SAM was not perfect. Sometimes the machine served as a pinch hitter, and sometimes it should have been benched. Overall, the numbers were good, but the quality was not. Yet he remained eager to try SAM again and see where the learning curve had left his crew. Within a month, WASCO began looking for another SAM job.

Scott and Zak both found this reception hard to stomach. SAM had hit some huge, unprecedented numbers—and had done so beside two untrained apprentices—but still wasn't good enough. To Scott, his machine seemed "definitely viable," but then he began to consider the counterarguments. He remembered the long list of fixes his engineers were tackling. And he admitted that masons couldn't run SAM on their own, which was the ultimate measure.

Pondering this latest facet only irritated Scott further, because while Stephen was reluctant to hand over SAM's tablet, John Lombardi continued to insist that SAM required two operators, when Scott didn't even want one. John's stance made Scott want to punch someone. It convinced him of the need to prove that SAM could get by on its own, without Stephen, or Ken, or any other CR chaperone.

19.

Seattle

Barely a month later, the job in Seattle presented a lot of firsts: John Lombardi's first time leading, Ken Carr's first time running SAM without Stephen at his side, and CR's first union gig. It was also SAM's first time running on the water—or, rather, beside the water of the Lake Washington Ship Canal. The building being bricked up was the Seattle Maritime Academy. From this corner of Seattle, beside the Ballard Bridge, just up from the locks and their fish ladders, it was hard to even imagine Laramie. It was CR's most scenic jobsite yet—not that anybody had time to enjoy it.

On account of torrential rains, the job had been delayed over a month, so there was pressure to get running. Cascade Masonry had 19,000 bricks to lay and had paid $10,000 to get SAM for eleven days. CR, prudently, promised 1,350 bricks per day. Math suggested the target was achievable in more than one way: If the crew rallied and kept SAM moving at 250bph, it would take under five and a half hours. More realistically, if the crew kept SAM moving at 180bph, it would take seven and a half hours, leaving just enough room for lunch.

SAM did not make a good impression on the first day. Setup took so long that the crew didn't start running until eleven a.m. And issues emerged from the get-go. Because the conveyer belt was loose, SAM placed some gun holes, which the men had to pull out and re-lay. Because the augur tube was loose, SAM buttered bricks inconsistently, and the wall ended up with tipped and shelved bricks. It already had a smile on the bottom. Somehow the machine ran away, and plowed into a mason, and placed a brick where there already was a brick. It slammed into itself. It refused to retarget, on account of an intermittent Wi-Fi signal—Wi-Fi, actually, was so poor that John Lombardi bought a long ethernet cable, figuring he'd tether the robot to the laser box so that at least the robot would have a chance. The day's tally came to only 675 bricks. The crew was not sold on SAM.

When Scott heard about this debacle, he made plans to head out to Seattle immediately with John Nolan, since it sounded like SAM's brain was fried. Tickets cost nearly a thousand dollars each. The men showed up at noon the next day, by which time SAM had laid 532 bricks without the tether. As sure as the Wi-Fi sucked, limiting SAM to 150bph, Scott reasoned that a tether would make his machine look incompetent.

The narrowness of the wall—only forty feet wide—was also not helping. The Hydro-Mobile seemed crowded. After lunch, when the men prepared to resume, they realized that the braces anchoring the scaffold to the building's frame were in the way, protruding so far that they blocked SAM's narrow track. They laid a few courses by hand, then lowered the deck so that they could reposition the whole scaffold. At the ground, they realized that the outriggers supporting the planks were already extended as far as they could be, so they were stuck. Scott asked Ron Adams, Cascade's owner, if it was possible to punch through the wall and anchor the Hydro-Mobile's towers, as was typically done, to the frame behind the brick veneer. But Ron—who was young but not *that* young—didn't like the idea. It would have entailed stepping on another union's turf.

The next day was even rougher. Scott assigned blame to man more than machine. The drama commenced when Scott noticed a mason measuring the wall slowly. Channeling Nate, Scott thought the man—all the men—could be working more efficiently. Two men should have been raising the story poles while two others continued readying. He told John Lombardi what he thought, and to pass the word on to the mason.

"I hear you," John said, "but I don't think we should say anything. Let him figure it out."

"That's silly," Scott said. "We know what works. Tell him."

John held his ground. Now annoyed, Scott instructed him to pass on the message. So John approached the mason and began, "Scott said you should—" Before he finished the statement, the mason stormed off, furious that some new guy—*some subcontractor!*—thought he was the boss of him. It was eight in the morning. The crew told these subcontractors to get off their Hydro-Mobile.

Sheepishly, Scott called Nate. "I think I just ruined our first union job," he said. Nate suggested that the mason was just making his point, and making it forcefully, as union men were wont to do. Scott next called Chris Raddell and began a long conversation about expectations.

At nine-thirty, the masonry crew found Ken and informed him that he had permission to return. He ascended, they measured, and another couple hours passed before SAM's first brick of the day went down. SAM's morning run rate remained below 100bph. The Wi-Fi signal was atrocious. That afternoon, someone nudged one of the poles, so all of SAM's bricks from then on went down tipped or shelved. Not that the number was so great. The day's total was 550.

Scott began to lose hope. He didn't think CR's plan was achievable and, as a result, wasn't sure how long SAM would stay in Seattle. Worse, he was not convinced that short setups were worth bothering over; he grew concerned about what this portended for the fast-approaching next job. He resumed the expectations conversation with Chris Raddell, who called the union masons "sensitive." Sensitivity aside, Scott was miffed that the masons hadn't anchored the scaffold as so many other firms had, miffed that the masons refused to stagger their lunches, miffed that the masons refused to use basket lifts to set the poles. To him, it didn't look like the masons were trying.

When Scott pulled up SAM's data that night, he found it painful to examine. Regarding Cascade's masons, the term "temperamental" came up. But Scott kept himself from acting impetuously. "We need to work with these guys to figure out how to be more efficient," he told his employees. This was no easy task, he said, "because of personality."

The next day, Scott and John Nolan flew back to New York. John Lombardi didn't show up at the jobsite, so it was just Ken. The Wi-Fi signal was so bad that it took an hour to lay one imperfect course. Ken developed a hypothesis that boat communication was interfering with their signal: The water was screwing them. The crew rolled SAM aside and finished the day using their own hands. For once, they lightened up, enjoyed themselves in exactly the way that Frank Gilbreth understood.

Friday's battle began with casting Scott's principle aside. Since there was no choice but to hardwire the ethernet cable to SAM, Ken was trying to waterproof the connection so that water could not immediately claim a second victory. It was all so strange, because only two months earlier, at yet another World of Concrete demo, that very machine had run fine.

So much was going wrong that, back in New York, Scott had a panic attack in his sleep. By the look of things, the machine in Seattle had some sort of concussion. That weekend, to decompress, he took his kids swimming.

An extended, distant scuba trip might have been wiser, because on Monday, Scott was around to follow the news from Seattle. It was an awful day. While setting up, one of the pole clamps broke; Ken drove to three stores to find a replacement part. After he'd returned and fixed the pole, the laser box on it went erratic. Instead of climbing to the proper course, it sank to the bottom of the pole, and then the laser ran away, as if sighting a brick across the canal. As a result, he wasn't ready to run SAM until two p.m. Once the machine started cooperating with the laser, it half-buttered a few bricks, pushed down too hard on others (so they had to be reset), and then crashed into a pillar. Ken couldn't tell if the map was off, or if the wall was curved, or if the camera on the gripper had been nudged. He ended up on the phone with Glenn, overwhelmed.

The masons felt the same way. Among other things, they decided that they had wasted a full morning, and that working with a trouble-ridden machine was far from ideal. "Screw this," they said. They voted not to work with SAM.

There was one more bit of news. John Lombardi, who had told Scott in Seattle, "This is not a good fit," didn't show up at the office in Victor.

That night, Scott called the day terrible, pathetic, a strikeout, a failure, and a total fucking disaster. "It's too bad Chris put so much into developing good relationships," he said. Because of all the farting around, CR stood on the edge of a precipice. Scott said the company's days on the Pacific were numbered. "I'm frustrated," he told me. "There are so many players in the Seattle market, and now we need to think about recovering.

"I'd like to swear up and down at Ken for not getting shit done," Scott said, but he knew that wouldn't be fair. In fact, he was partly grateful for what Ken had done: He'd exposed a rash of new bugs, and Scott planned to keep a cool head and engineer his way out of them. In a way, this was what he had asked for with a minimum viable product. The timing sucked, for sure, and Eric Ries had perhaps understated the pain associated with his approach. If anything, Scott was, just like Zak, annoyed at Stephen. For nearly a year, Stephen's competence had masked many of SAM's issues. But without Stephen, CR wouldn't have had numbers to show off—so it was hard to be mad at him. He'd gotten *bricks in the wall.* Scott realized his irritation lay not with any person but with his situation.

Scott was also annoyed with the union, whose posturing stung of hypocrisy. It was too much for Scott that union bricklayers claimed they wanted to work with a machine but grew irritated when it didn't run perfectly—if it laid bricks perfectly, 100 percent of the time, human bricklayers would be moot. Wasn't that what they wanted to avoid? Scott had always known that trying to get a *completely* autonomous robot to perform at a construction site would be, in the phrasing of his own patent, "a prescription for disaster," but he hadn't foreseen that creating a cooperative *semi*-autonomous robot would present no less of a challenge. He was flummoxed.

He was annoyed at himself as well. He had thought CR had a good product. He had thought they could achieve so much. He had thought the Seattle job was a no-brainer. But for the first time, CR's operator had stepped back and handed off the reins. "Clearly, they can't run the machine," Scott said. "Ken can barely run the machine." He thought about Ken. "From a personality standpoint, he's doing fine. He has the right attitude, capability." Ken had put a lot of energy into diplomacy with the masons. But CR had done itself no favors.

His company should have provided more guidance, Scott allowed. That was what he and John Lombardi had disagreed most about. John Lombardi, who knew construction, had wanted to let Cascade figure things out. Scott, the process engineer, had wanted to teach the most efficient process. He wasn't sure if John didn't fully understand the state of SAM's technology—but it mattered little, because he was gone.

"If I had more control," Scott said, "I wouldn't have to deal with this shit." It boiled down to control. "We give 'em Steve, otherwise we get bit in the ass." He couldn't see trials working any other way.

"It leaves us two options," he said. "One, we become a masonry company, or two, we own everything." He hinted of an invading military: "We kick ass, knock out a wall in two weeks, and if we do it, they figure they can, too—after a few weeks of learning.

"We fed 'em to the wolves," he said. As his old swim coach had realized, the biggest mistake could become the biggest learning experience—but Scott wasn't ready to think positively yet.

"We're in a tough spot," he said. It was CR's lowest spot so far, he thought—so low it made Tennessee look like a home run. "It's a back-and-forth yo-yo of shit quality and good production, like in Tennessee, or shit production and good quality, like in Seattle." How desperately he wanted to escape the pull of these two planets.

He thought of a faraway land, a land in Missouri, where CR's next job was coming together. He was optimistic because it included a long wall, and because Stephen would, at least for a time, be there to demonstrate and help. Scott also had faith in the mason. "We should have big success. And if I could have one job to have success on, that's it. If Seattle totally fails, everyone in the union will be talking about it. But if we knock it out of the park in St. Louis, we'll have divergent stories."

The uncertainty of divergent stories was now the best Scott could hope for. Don Golini had warned him about such a situation.

On the seventh day in Seattle, SAM laid 617 bricks but punched through four of them. The men called SAM stupid. They were done futzing with the big beast, even on a leash.

When Scott assessed the day with Ken, Scott was testy, trying to figure out how the job had gone south so fast. Nearly a year after Lunenburg, CR's worst job among many bad ones, a job had gone more disastrously. He was pissed because SAM's utilization rate stood at 40 percent—and he believed Ken should have been better. "I think we gotta get you back here and get more training and re-create some issues so our engineers can understand what the hell is going on," he told Ken. He wanted Ken confident, full of experience, capable of hitting fifteen hundred bricks.

Scott called Ron Adams, Cascade's owner. He wanted to apologize on behalf of a shot flux capacitor, but he knew Ron wouldn't buy it. Ron had lost money. SAM hadn't laid three thousand bricks.

They talked, and Scott salvaged an amicable truce. He told Ron that the Seattle numbers were the worst CR had seen, which suggested there was something wrong with SAM. He said he could fix every problem in the next three months, and he knew they could do better in every regard. "Clearly, we didn't perform," he said, "but we want to continue to work with you, and I'd like you to be an adviser. Give us advice from your team, tell us how we can do better. Then come out to Rochester in three months and see improvements, and we can talk about future jobs once you see it for yourself."

Scott pulled this Eric Ries move out of his ass, because he was determined to end on a positive note. It worked. Ron agreed, and Scott even offered him his money back, in the form of a credit on a subsequent job. He explained his thinking this way: "I can't have Ron frustrated and go out and badmouth us. I want him to defend us. I want him to say, *No way, buddy, I learned what it's about, and we're gonna do it again in a few months.*"

Even though it was April Fools' Day 2016, Scott sent an email to everyone with a priority list attached. The list, he wrote, represented the things that needed doing to have a customer-ready machine. This made it "the most important list in the company." Most items on the list had dates assigned to them, and none was later than June 1. "I cannot stress enough how critical this list is to the importance of this company," Scott wrote. "It's not only critical that we get it done, but that we get it done fast." He saw it as the key to their success.

On Monday in a meeting, Scott decided more bluntness was needed. "We had a rough start out in Seattle last week," he told the staff, understating the situation. Cascade's owner, he said, had questioned when they'd be able to run SAM. The bottom line, Scott said, was "we don't have a robust machine that can be run by masons in the field. We're Band-Aiding it in the field, so it's of utmost importance that this list gets resolved . . . I want to review it regularly." His list of priorities was not short.

The engineers were silent. How Scott remained so motivated about an unlikely future mystified them. Their second construction season was upon them, and their calendar was nearly empty. Yet somehow their boss saw opportunity. He wanted to avert another debacle by fixing the machine before attacking any job.

Scott admitted many items were already on his priority list, but some—like Wi-Fi—had risen to the top. He also hated the tether Band-Aid and wanted a real fix. "Everything on this list is focused on quality or user experience," he said. Across the top of the list, there were forty-four words, which included the statement "a customer should be able to easily run SAM" and the phrase "consistent, robust results."

To John Nolan fell laser, Wi-Fi, and pump problems. To Mike Oklevitch, mortaring problems. To Chris Johnson, dynamic compensation. To Kim Heng, a wider gripper and a new plate. To Tim Voorheis, frame problems (stiffness and clearance), mixer problems (overheating), augur problems (broken bearings), and conveyer problems (flipped bricks).

John was the first to pipe up. Fixing the conveyer so that it didn't intermittently flip bricks, he said, was such a big project that it made more sense to just adjust bricks by hand. It was called quality control. Scott agreed that the project had scope, but he insisted on an engineering plan. "Dealing with a gun hole every four bricks is not a solution," he said.

Chris Raddell wanted to give customers a troubleshooting chart, which riled John even more. "If we're able to write descriptive things about problems, we should be able to solve them," he said. He had a point. "For most, you hit reset or power-cycle the machine," John said.

"It's not worth it to write: 'We're not really sure what's wrong, just hit reset. That's pretty much it. There's not much else you can do.'"

Scott got ahead of himself and mentioned the laser tracker he and his engineers had long dreamed of. They were finally looking into it. He was sure it would improve SAM but deemed the thing below the level of a requirement. "Right," John said, "'cause it's not gonna be ready." Tim had concerns about realistic schedules. Putting a pivot axle on SAM's frame (to reduce wobbling) would take two months. Scott, for his part, pushed back. He thought some of the due dates on the list—like the one for improved dynamic compensation—were made up. "Guys," he said. "Two months."

A few days later, Scott maintained the pressure. On speakerphone, he told his engineers, "We need good testing. With issues we're running into, I want to validate. I don't want to put shit in the field anymore. Let's validate. Let's get on the Hydro. Full runs. No more fake mortar." He was running through the list.

Already, Mike Oklevitch had developed a new buttering path that laid mortar better and happened to save a third of a second. Bricks on this path came within a quarter inch of SAM's frame—but Mike Oklevitch thought this clearance was okay. Tim had built a new laser box, with an improved gearbox and a firmer connection to the smart pole, and was testing it. He was looking at the wiggle in the augur tube and thought the mixer motor needed a waterproof cover before that became a headache. John was hardwiring a new controller to the laser box so that Scott wouldn't have to deal with a tether. Glenn, who had looked into improving the map tool, said doing so required redesigning its back-end database so that it could run quickly and smoothly. Scott didn't like moving backward, but he listened to Glenn. Kim had designed a new fold-down track and was having a ten-foot segment built for testing. He thought it would work, but he wasn't ready to order hundreds of feet of the stuff. This made Scott grumpy, because he wanted to get the track out ASAP—anything to improve setup time and relieve headaches. Chris Johnson, who'd had improved dynamic compensation on his plate for nine months, was waiting for the manufacturer of a particular momentum-measuring sensor to respond to his calls. He said the company seemed to have dropped off the face of the earth. Once he got the part, he said, he should know by the end of the month if the approach would work.

"Hang on," Scott said. Given his time line, this wishy-washy statement made him want to swear.

Tim defended Chris Johnson. "We need to validate for full implementation," he said.

"This seems like a never-ending project," Scott said.

"It's a resourcing thing," Tim countered. "It always is."

They were putting out too many fires.

That month, Scott's stress rose so high that Torrey told him she was concerned about his blood pressure causing a heart attack.

Scott was confounded by a lot: by the variability between the different machines and different jobsites; that SAM had not yet attained the speed of his dreams; and that his engineers had not remedied the intermittency of the mixer motor, which manifested in the form of shitty mortaring, which might have been the only thing that displeased masons more than OSHA.

He was confounded even more that SAM was not yet customer-ready. In Missouri, Scott had tried handing SAM off to a young man who'd spent months learning how to run the machine, even received tips straight from SAM-guru Stephen Kean. *Listen to SAM*, Stephen had said. *If you don't hear the Stäubli, and the blue light's flashing, that's no bricks. If the pushing ram keeps trying, that's a double feed. Watch the bottom corner of the tablet. Read the wall—you can tell if the laser's off from the last few bricks.* One morning, without Stephen by his side (but nonetheless on-site), this young man had gotten in twelve hundred pewter bricks on the new wing of the medical center before noon. Scott had heard the news and was thrilled; it was a position he'd aspired to be in for years. He called it monumental. In an excited state, he'd summoned Stephen back to Victor. Then everything had fallen apart.

Three hundred bricks into the day, SAM pooped mortar for a minute straight while placing six bricks. Soon after, the laser fell under some kind of spell and vanished. After that, the Stäubli froze and refused to let go of a brick. In all, the poor man endured over three hours of downtime, which amounted to a thousand bricks going nowhere. Alarmed, over his head, and exhausted, the dazed operator had spent half his day on the phone with nearly every engineer at Construction Robotics, seeking emergency assistance. He'd spoken with Stephen fourteen times; the last time, he'd said, "You're never allowed to leave again!"

If there was any good news, it was that SAM was faster than ever. By May, Chris Johnson reported a cycle time of 8.7 seconds, which amounted to more than 400bph. Scott called it "exactly where we wanna be" and surprised his engineers by briefly revealing a trait he rarely displayed: satisfaction.

But the speed presented its own problem. SAM's arm moved so fast that when it rotated while rolling in one direction, the mud that had so carefully been applied to bricks started flinging off. Mike Oklevitch had, off and on, noticed the problem for most of a year, but John Nolan had always disputed it, and in any case, it had never risen to the level of problematic. To be clear, the final rotation of a buttered brick upon placement was always clockwise, but the amount of rotation changed depending which way SAM was moving. Because the Stäubli was a left arm, and because the nozzle pointed up (no point in fighting gravity), SAM always buttered along the same path: from a brick's right head, to its bed, to its left head. When SAM was moving to the right and placing bricks down and to the left, it started the buttering sequence on the second step, with the bed, and then squished mortar onto the left head. When SAM was moving to the left and placing bricks down and to the right, the machine buttered the right head and then the bed—and it was these bricks that could not handle the spin. On these bricks, the pump couldn't start fast enough to apply mortar to the short head with sufficient force. Glenn provided a quick fix to the problem by writing code that made SAM run in only one direction—like a typewriter instead of a printer—but this, not surprisingly, did not sit well with Scott.

Soon enough, Mike Oklevitch persuaded Glenn and Chris Johnson and John Nolan that his new buttering path, if synchronized with the nozzle and the mortar pump, kept the mud on and didn't sacrifice any speed. The solution hinged largely on tilting the nozzle out at a 45-degree angle, an advancement they henceforth hailed as 45-degree buttering (FDB). FDB was tricky—it required redesigning the sensor on the back of SAM's gripper and cutting out a small section of SAM's frame—but it was a huge leap. It allowed for a refined motion that increased SAM's speed by *eighty bricks per hour* and allowed Mike Oklevitch to focus not merely on the particulars of mortar makeup but on its universal application. If ever CR had a prime example of learning and iterating, FDB was it.

By the time FDB was developed, Stephen had returned to Missouri to help run the machine, which looked battle-scarred. Among other injuries, it had a small tear in the conveyer belt, the hair dryer strapped on with zip ties, a camera lens covered with a sandwich bag and held in place with a rubber band, and a cabinet door braced open with a cam strap. Still, it placed bricks quickly. Where the gripper used to hum for one second upon placement, to vibrate the mud, now it hummed for a fraction of the time. The machine sounded like it was in a hurry.

Spring was a volatile season. That time of year, water could fall frozen in spheres the size of tennis balls, or it could form funnels, travel uphill, carry things away. To Scott, it was a time of transition, a tipping point for the scrappy little start-up. In a serious voice, he told his engineers that he intended to get shit done. His focus, he said, was to make sure SAM and its eleventh software edition were stable. He wanted 110 percent confidence from hours of testing. "Test the crap out of it," he said. He wanted machines in tip-top shape. "Thinking we can build two-hundred-brick walls and call it good," he told his engineers, "is bullshit." He wanted his engineers to repeat the situations where bugs emerged, isolate them, determine if they were software- or Wi-Fi-related, nix them, and then double-check that the bugs had been nixed.

No matter what, he didn't want another Seattle. But he didn't want to be overcautious, either—which was why the tether scheme up in the Puget Sound annoyed him. He was sure no mason would ever find SAM appealing tethered. The whole point, he said, was to be able to say, *Look at our amazing product. Don't you want to buy it?*

He pushed the engineers until John and Tim were busier than ever. The director of engineering, meanwhile—who was trying to pin down an official list of every component in SAM—struggled to keep up with the rate of change. John put the situation in context for him: "It's what we always do. We push forward even though we're too busy."

But Scott's timeline in spring 2016 wasn't arbitrary. He wanted Chris Johnson to hurry up and stabilize v1.11 so that he could use it for a demo with a local TV station. He wanted to hit twenty-four hundred bricks in an eight-hour day. That was his magic number. It's worth noting that this was twice the number that, not so long before, he'd vowed to offer a bonus for. SAM's new cycle time made Scott anxious, almost desperate to get on a job. He wanted to hear a mason contractor, in awe of four hundred bricks per hour, say, *Holy shit, how do I get one of these things?*

Yet Scott wanted more than speed. He wanted the machine easy to use, especially the Stäubli arm. When the arm reached too high or too far, or otherwise approached the limits of its joints, it froze, as if turned to stone. Unsticking a stuck arm required pecking on a small gray universally despised keypad. Its buttons might have been in Klingon, for all the loathing it summoned. Scott wanted his engineers to transfer all the Stäubli's controls onto the iPad (otherwise known as the HMI, for human/machine interface), which was less intimidating. And he wanted his engineers to mock up a visor, so a certain angle of sunlight wouldn't continue to trip up the laser camera on the gripper. He also wanted them to remedy SAM's low-rider problem, since sometimes the machine got beached. It happened when the machine's forklift pockets scraped over seams in ultra-wide Hydro-Mobile decks—and it was always embarrassing, like a bulldozer stymied by an anthill. Half a million dollars of modern automated-bricklaying technology beached like a big blubbery animal. He wanted Chris Johnson and Glenn to solve dynamic compensation—he'd been waiting six months now. He wanted more than he asked for, knowing that it was not the best time to also have his engineers devise a camera package that scanned a wall and automatically created a map, sans human measuring.

Scott wanted one more thing. He wanted to refine the poles so that setting them up would take under thirty minutes. He wanted his guys

to practice pole raises with new brackets that Stephen had thoughtfully designed. He was even open to using shorter poles, though they'd limit the number of bricks a mason could lay in one position. Scott said he wanted to watch and time his employees. "Micromanage it?" Tim asked. "Yeah," Scott said. It had come to that.

In the machine he'd been building for a decade, Scott wanted the reliability of a tractor or lawn mower. He'd already broken nine seconds per brick, Stäubli's fundamental barrier—but his engineers seemed willing to glom more parts onto SAM in ugly, hokey, unsatisfactory compromises. He still dreamed of something simple and elegant, beyond the realm of anything minimally viable. Basically, Scott was tired of the MVP approach, and it showed. "It has to just fire up and go," he said. "I need to be able to do it myself. I want to say, *It was easy. I had five hundred bricks before breakfast.*"

This last notion so surprised him the moment it came out of his mouth that he repeated it. "That could be our slogan," Scott said. *"Five hundred bricks before breakfast."*

20.

Double

As summer began, the engineers attacked the machine as they had the Alpha, because SAM was still unhealthy. Tim addressed everything he could. He called it multiple surgeries at one time—brain and leg. After Scott suffered double feeds all day on a demo, Tim examined SAM's pusher bar but got pulled away from the repair four times for higher-priority repairs. In short, it was triage. He installed shorter forklift pockets on all the machines—they had half a dozen—except one. He considered ways to block sunlight from hitting and interfering with SAM's camera. He examined the new clamps that Stephen had designed to quicken pole setups. When asked for a solution to the pump-runaway bug, Tim suggested, "Carry a shovel?"

Meanwhile, Kim designed and tested a new gripper. Kerry pondered compressor fixes. Mike Oklevitch refined the way that the machine calibrated mortar, correlating the current drawn by the mixer with its slump. He also minded the gap, because he didn't want every brick rubbing against his precious nozzle. The engineers consulted Wi-Fi experts. They tweaked the UI. Chris Johnson reported a cycle time of 8.3 seconds. Regarding the wandering laser, which Scott desperately wanted fixed, they might as well have consulted an oracle.

Leica's laser tracker (borrowed, not purchased) finally arrived, and after learning to "talk" with it, Glenn began to test its speed and robustness to see if it could deal with their Wi-Fi network, and if it could work in the sun.

This was done, however, without John Nolan. He had resigned. The announcement was a big shock to Scott and registered as the worst possible loss. John, who had replaced Erwin and had worked alongside Rocky, had designed SAM's brain over the previous two years. His personality—his resistance and honesty—had guided so much of CR's course, even if he'd needled Zak for most of his tenure. John Nolan, whose background included fuel cells and machine making, became the eighth person to leave Construction Robotics. He was going to Genesee Brewing—to turn water into beer.

And then WASCO, the Tennessee firm, reentered the scene. So did Arch Masonry, in Pennsylvania. Arch, which booked SAM for a one-month rental, told Scott that if they liked what they saw, they would lease SAM for longer—and if they still liked it, they would snap up a machine. They were looking for something worthy. Starting July 13, when Stephen turned the key in Nashville (Ken had a week's head start), Scott got what he'd long wanted: two machines, at two different jobs, running simultaneously. The double play lasted for seven weeks, which was long enough for everyone involved.

The job in Tennessee, just northeast of Nashville, was in a big field, and the job in Pennsylvania, just north of Pittsburgh, was at the intersection of two interstate highways. On the former, the bricklaying machine (deemed SAM4) sat before a quaint Baptist college, and at the latter, the bricklaying machine (SAM2) addressed a large chain hotel.

During the first week, neither Ken nor Stephen got much more than two hours of daily run time. In Tennessee, the problem rested in the heavens: Thunderstorms rolled in, driving WASCO's crew home. In Pennsylvania, the problem was the bricks, and the mud, and Arch's crew. Covered in dust, the bricks didn't slide properly off of the conveyer, resulting in double feeds and gun holes. The mud was too thin. As for the crew, the mason couldn't digest the map tool, and the telehandler operator couldn't be found or bothered or both—which meant delays delivering bricks and mortar. As a result, everything took longer than planned. Where Stephen cranked out 780 bricks in a day, at 330bph, Ken ran 10 to 15 percent slower.

A heat dome settled over the country during the second week, resulting in the greatest scorchers of the year. Two hard hats in Missouri died from the heat. Stephen, soaked from glasses to boots, thought conditions rivaled those from the Lab School job, which put them at satanic levels. It was unbearably hot (and unbearably unpleasant weather for heavy boots and pants). Even the water in his hotel's pool was tepid. Chris Johnson worried that the heat would be the undoing of SAM's compressor, but Stephen thought pointing the fan at it would be defense enough. He was right—and the compressor on Ken's machine overheated first. Other problems presented themselves to Ken as well. His iPad dropped his measurements four times in a row and wouldn't stay charged. His generator crapped out. His crew kept changing, such that he was stuck with SAM newbies daily, and he couldn't reach the foreman: The man wouldn't answer his phone. On top of that, the Hydro-Mobile was wonky, and this new crew couldn't find the necessary parts to fix it. Scott heard this and called the situation in Pittsburgh a clusterfuck. Eager for a thousand-brick day, he had Zak put together a document outlining the steps necessary to get there. A road map, as it were. Scott told Ken to sit down with the foreman and go over it. He told Ken to *regroup*.

But for the heat, Stephen was feeling pretty good. Under his watch, SAM placed 1,128 bricks in a day, for a total (including hand-laid bricks) of over 1,600. He predicted that he'd be able to hit the 1,500 count that WASCO was looking for daily. Better still, the crew was already beginning to run SAM. They were good workers, and they liked the machine. Their boss was happy. When Stephen heard of Ken's headaches, he mentioned that he hadn't needed to touch the Stäubli keypad once and advised his Pennsylvania colleague and onetime tutee not to overthink details.

Ken said he, too, hadn't touched the Stäubli keypad. The problems were not in SAM's arm but everywhere else. He stumbled along as he dodged a camera crew on the deck of his scaffold and tried to make do with too-thin mud. It was splattering the wall, and he was pretty sure it had left a smile as a result of settling courses. One day, in addition to the usual delay incurred by setting up poles, Ken faced a trickier dilemma: a rumor. Word spread that OSHA was on the way for an inspection, so everyone ran off to make the site shipshape and Bristol fashion. SAM was abandoned. Ken's daily brick count remained under four hundred for two days in a row. Hearing this, Scott reminded Ken to stick to Zak's road map. Because Scott still thought he could convince Arch to use SAM for three months and maybe longer, he headed south to assist yet again.

Fed up with slow pole setups, Scott started early, ditched the sucky twelve-footers, and went with eight-footers instead. Finally, SAM got some real run time, until a morning thunderstorm scared everyone off the scaffold. At Ken's hand, SAM had knocked in 750 bricks in under three hours, but the storm dragged on for nearly as long. At last, when conditions returned to buildable, Scott climbed up the Hydro-Mobile and discovered that SAM's mixer was not mixing. He suspected that one of the crew—inadvertently or not—had turned it off. It was not quite lunchtime. When he turned the mixer on, it didn't move. The thing was seized; it was the worst mortar seizure Scott had seen. As luck had it, Scott had arranged for the head of Arch to swing by after lunch and check out SAM. Scott thought: *Oh, fuck.*

Since the crew had gone to lunch, it fell on Zak (who was on-site with his brother-in-law to take marketing photos) to get digging. Barehanded, he pulled out globs of cold lava, bloodying his hands. He grabbed a brick tie and stabbed at the hard stuff, breaking it into clumps, scraping it off the sides of the hopper. Mike Oklevitch, who had also come to help for the day, pulled the nozzle off the augur tube and then helped Scott wrench on the augur itself. All told, two engineers, an auto mechanic, and a business degree spent forty minutes fighting mud. Finally, they cleared enough crud from the machine that the augur was able to turn. On schedule, the head of Arch Masonry showed up.

Scott told the boss what had transpired and explained that it was an atypical and preventable problem. He assumed it was an accident and said he'd train Arch's guys not to let it happen again. The boss told Scott that he was unhappy, because seven or eight hundred bricks a day could not yield enough profit to justify SAM's use. He wanted fifteen hundred, and that day's brick count was the first on the job to exceed one thousand. "We gotta do something," he said. Scott said he very much wanted to hit those numbers, but doing so would take a trained crew attending to SAM. He said he wanted to make a game plan and stick to it on the next wall. (The last time he'd made a game plan was on the Lunenburg job, which was not a good sign.)

For the first time, Scott began to loathe the rain. Ken had been having a great day and was teed up to have the best day yet—and then water fell from the sky. As a result, the day's production was mediocre. It was the tenth day on the job, and SAM hadn't run longer than three hours in a day. Its daily average stood at 525 bricks.

Stephen, meanwhile, got SAM to lay down 938 bricks. The day before was 864. The machine, he said, was running well.

Enlightenment arrived in the third week. Scott realized the size of the wall in Pittsburgh was inhibiting production. It was too narrow, like the Laramie job. A thousand bricks was nearly impossible, as it would take setting up the poles three times. Yet when Ken compelled SAM to lay 835, Arch's boss was ecstatic. He counted on 300 bricks per man, and anything beyond stood out. His men, though, remained unthrilled. They didn't like SAM. SAM was fussy, and the machine's safety relays kept tripping for no reason, stopping the machine midaction. Stephen suspected that water had seeped into one of the bumpers on Ken's machine, shorting it. Not that it mattered much. The Hydro-Mobile was broken, and so was the telehandler. They couldn't move up, and they couldn't get mud. Now Arch needed to regroup. Since it didn't make sense for Ken to sit there and do nothing, he returned to Rochester and waited for Arch to rejigger.

On Stephen's tenth day, his daily average stood at seven hundred bricks. He was on a tricky wall, with lots of windows and piers, but he was hustling. He wished he had a big wall. One day, the compressor overheated twice. Two days later, it overheated again—that time, he smelled plastic burning. The following day, the generator crapped out completely. To a Cummins dealer he went.

The fourth week began with bad news on both sides of the Mason-Dixon Line. To the south, the Cummins dealer declared that fixing the generator would take a week, since the necessary part, in the words of Wallace Stevens, was like nothing else in Tennessee. To the north, the mixer belt snapped before Ken got four hundred bricks down. When he got to a hardware store, he bought not just a replacement belt but a backup as well. He returned to the site, installed the belt, and turned on the machine to test it. Then the regulator on the generator broke. Now Scott had two machines not running on account of generator trouble. Ken ran to Home Depot, found closed doors, and then ran to Walmart, where he found the part he needed. He returned to the jobsite and wrenched until two-thirty a.m. At sunup the next day, the machine hummed to life.

Stephen waited for a generator to arrive from Victor. With the new generator, he sprang back to action quickly. He got 770, then above the windows got 1,071, and then 1,134. He topped out the wall, earning himself a two-week break until the next wall was ready.

With a new belt, Ken fared less well. The wall was so narrow that SAM could place only a dozen bricks before the mason had to place a few at the end. And the mason always placed those few bricks with a certain arrogance. When quitting time neared, the mason and his crew split pronto-speed—sans cleanup, sans setting up for the next day, sans regrouping. For four days straight, Ken didn't break five hundred.

Over the next two weeks, the situation in Pennsylvania unraveled. The crew tried to run SAM, but the Wi-Fi wouldn't connect, so they shut the machine down. It was as if Ken were cursed. The crew ended up hand-laying bricks casually, finishing the day with 725. They were convinced SAM was a dud. While they were moving the dud to the side, SAM collided with one of the masons. It wasn't a high-speed collision, and he wasn't run over, but he got a good poke in the hip from the conveyer's unbumpered sharp corner. It left a bruise so deep and sore that the mason showed it to a doctor. (*"Doc, a robot hit me."*) SAM's compressor overheated again, and the belt broke again. The mixer nearly seized up another time. Scott, of course, knew of Ken's heroic efforts and offered him a few days off upon his return. But there was still a wall to build.

Ken swung through Arch's office and heard the estimator call SAM a joke. But the crew really vexed Ken. The laborer barely thought three seconds ahead. He connected the wrong propane tank to SAM, rupturing the regulator. One of the masons was downright belligerent and seemed to be itching for a fight. Another seemed intent on blocking the laser all day, no matter how loudly SAM chirped. Someone on the crew covered up the security cameras, in case their superintendent had plans to spy on them. Two foremen quit. Ken felt like a babysitter. Scott, hearing this, called Arch's men a "yo-yo crew." The yo-yos hated SAM. The crew's lone SAM fan asked if Ken could get the beast off the scaffold, so that he and his colleagues could lay by hand.

Yet during that time, Arch posted a promotional video on its website. In thirty seconds, split screens showed the modern mason's machines: a brick saw, a telehandler, a Hydro-Mobile, and SAM. "Arch Masonry," a narrating voice said, "the area's leading contractor, sparing no expense to ensure that all the latest equipment is on the job." Footage cut to an overhead shot of SAM and then a close-up of the machine's big red arm. The voice continued: "And that includes the use of innovative robotics, combining old-world craftsmanship with modern technology . . ." How strange that even those who despised new technology felt compelled to hail it.

Arch's boss showed up one day, and SAM and the men together put down fifteen hundred bricks. The guys knew not to lollygag in front of their boss. Inspired, Scott drove down to Pittsburgh and had lunch with the boss.

"We proved it," Scott said.

"You kinda proved it," the boss said.

In the last week, the crew's wiliness became too much for Ken to bear. He was exasperated. The men, in his words, were miserable and slow and full of excuses. He wanted to be done and wondered if it might be wisest to finish alone, at night. The last day arrived, and Ken wished he'd followed his intuition. After delays and delays, SAM laid a single brick—at which point one of the crew said, *Enough of this fucking shit.* They permitted SAM one miserable course and then neglected the machine. The mixer seized up, and Ken spent the rest of the day chipping it out. When he looked up, the crew had vanished. He considered hot-wiring the telehandler, moving SAM himself, and hightailing it out of Pennsylvania.

Stephen, meanwhile, had resumed on a big flat wall, with 135 bricks per course. While Ken chipped away, Stephen laid 1,685 SAM bricks. The next day he got 1,586. The day after that, he got 1,600 by eleven-thirty a.m., then raised the poles and finished with 1,977 bricks. He wrapped up the job with two more days over a thousand.

During the last two weeks of August, Stephen's average daily production was 1,100. In two months, he laid 18,000 bricks. Ken, five hundred miles northeast, laid half of that.

Soon after, when Scott asked both Arch and WASCO if they wanted to buy SAM, both said they were not ready to jump.

A better question was: Why didn't Scott's field-service technicians jump ship? Why, given all the suffering and struggle, did they stay? Because they believed in Scott and his machine. Because they were getting close to doing something nobody had ever done before. Because even if work often demanded exceptionally long days, and weekends, and the various miseries and indignities of construction sites, there was, thanks to Scott, a zero-bullshit factor about their toil. They had no obligation to write reports, or sit through sales meetings, or deal with nincompoop middle managers. To overcome the challenges before them, they were free to nerd out as much as they wanted—in fact, the more they did so, the more they had their boss's ear. Things were moving quickly, which was not the case with so many companies, large or small, as they well knew. Where else was work so rewarding, so exciting, so real?

In fact, Stephen was far less concerned with his own comfort than with that of his successor. When Scott began looking for a third field-service technician—a requirement if CR was to grow—Stephen thought that Scott was foolish not to insist on candidates with automotive experience. Without automotive experience, Stephen thought, someone operating SAM would not prevail through thick and thin.

That Scott's most talented operator doubted Scott's ability to find qualified staff said something about the quality of the product he'd developed.

21.

Wisconsin

Where CR's engineers had only months before cranked out huge efficiency gains—via a new buttering path and pole setup—they were, by the end of the summer, polishing SAM, refining the machine. The tablet was improved. The track was slicker. Via 4G cellular, SAM's data uploaded automatically (which allowed Scott to focus more obsessively on numbers). As a result, Stephen found that he could sometimes, as he phrased it, peace out.

This was good, not just because Caterpillar and John Deere were keeping tabs on Construction Robotics but because the plan in Wisconsin was certainly ambitious. J. H. Findorff & Son wanted to erect eleven walls in eleven days. The walls on the elementary school were short and didn't require scaffold anchors—but still, Findorff's agenda was a bit like setting off from Lees Ferry with only eleven days of food. The math presented itself quickly: You had to put twenty-one miles of the Colorado River behind you every day. Or else.

Findorff was a Madison-based construction company so impressive that it was almost immeasurable, with a large and stunning headquarters. A just completed renovation showed off every trade: In equal parts, the structure paraded steel, glass, wood, concrete, and bricks. It was ravishing, really—with a cavernous foyer, a kitchen, a workout room, a locker room, a lounge, and a bar, as well as a parking garage and a "war room" where project estimates were compiled. Its roof did not leak. From nearly anywhere in the building, one had expansive views of Lake Monona, whose surface rippled and sparkled differently every day.

Scott and Nate had met some of Findorff's higher-ups in April and concluded quickly that Findorff was thriving. The company employed 500 people—750 in the summer—and performed almost half a billion dollars of construction every year. Because the company had been around since 1890, it had built many of the hospitals and schools in town, as well as the county courthouse and the state capitol. The company had built so much, in fact, that its motto was "Builders of Madison." Nate and Scott were wowed to such a degree that they left wondering why their own hometowns of Syracuse and Rochester weren't booming the way Madison was.

Nate was even beyond wowed. He sensed a special attitude at Findorff, one he'd been seeking out nearly forever. It was invigorating—like, he said, "a shot of salt in my body." He called Findorff not merely remarkable but the best company he'd talked to in his life, so alluring that he didn't want to return to Hueber-Breuer. He said he'd work at Findorff for free, and meant it. He was sure that the day CR met Findorff was the best day in CR's history; sure it followed from the luck that Don Golini had insisted on holding out for; sure that Findorff was the company through which the world would learn of SAM.

Before the job began, Findorff's foreman visited Victor twice and got comfortable running SAM. His crew had a great attitude, free of the traits that so defined bricklayers. Scott, for his part, tried to match that attitude and insisted that his engineers go over SAM thoroughly with a preflight checklist. He wanted all of CR to behave proactively, as if the company's life depended on the machine's next flight.

Findorff got a slow start because SAM wouldn't put bricks down where it was supposed to. Ken, who had somehow been saddled with another difficult job, made some adjustments and checked the first wall. He made more adjustments and checked again. He was about to adjust the machine a third time when he realized one of the poles had been clamped to a big piece of wood that was waving in the wind. No wonder. Go figure that at Eagle Point, in DeForest, Wisconsin, the wind blew. The Findorff team eked out six hundred bricks from SAM, and the quality of the wall did not sit well with them. The bricks were awfully banana-shaped, but still.

Scott was on-site, too, coordinating. Among other things, he wanted to push the latest software into this machine, but he delayed because of the well-understood risk. He held back all week: That was planning ahead.

When he and Ken called Victor with the first day's update, Stephen suggested moving the wobbly pole in, closer to the center of the wall. The three of them knew this portended losing a few SAM bricks on every course, but Stephen thought it was preferable to yield some bricks to men lest they become 100 percent dependent on SAM. Everyone knew the machine was far from perfect, and full dependence would just put it under more scrutiny. Plus, one could get more bricks by combining man and machine. Scott did not like this perspective, because his goal, as always, was to show big machine numbers. "I understand your point," he said, "but we should talk more about it." This was his new longhand for "noted."

The Findorff crew had set up a second Hydro-Mobile on the next wall, and they leapfrogged SAM onto it the next day. This scaffold, though, had a wonky quality: The track seemed loose. Also, it wasn't quite in the right spot, so they pulled SAM off and repositioned it. The crew wasn't thrilled, but they weren't pissed off, either. Finally, they got SAM going. The crew insisted on operating the machine, so Ken stood by in case he was needed. The crew impressed him; they were neat and clean and organized, and they talked to each other. They behaved like big boys. It was only day two, and they cranked out more than eight hundred bricks, at an average of 10.2 seconds each, before rain arrived. Scott, that afternoon, said he was not dissatisfied.

The crew yo-yoed SAM onto the third setup, in front of the third wall. This was all the yo-yoing they did: They picked and placed SAM on this Hydro-Mobile one day and that Hydro-Mobile the next. A metaphysicist might have pointed out that, in this act, the men were imitating the motion of the world's first robot: *Pick and place. Pick and place.*

Ken was relieved. But SAM—the same SAM that had just been in

Tennessee—had some yo-yo moves of its own. The laser box wouldn't move up a course. Then, under instruction, it blasted up the pole, as if headed for orbit, and got jammed ten feet up. Two courses from the top of the wall, the mixer went bonkers and spit out all of its contents, as if it had food poisoning. Ken called these two bugs "ghost errors," and while they left the crew mystified, they did not sour the day's progress. The men finished the little wall by hand.

The leapfrogging continued, and SAM ended up on a ninety-foot Hydro-Mobile, before the biggest wall of the job. A new batch of bricks arrived—and to Scott's contentment, they were gloriously square. The crew drove SAM to place seven hundred of them, and then they hit a stone lintel and stopped the robot. The stop aggravated Scott because of the lost opportunity. SAM sat there like a fat toad. Scott saw room for the masons to split the wall, let SAM run on one half while they set stone on the other, but this they did not do. It was their prerogative, and he'd learned better than to interfere.

On the last day of the first week, SAM laid fourteen hundred bricks but had developed a fifteen-second lag on every placement, which so annoyed one mason that he left. When he hit the gas, he wanted the car to go. *Now.* He did not want to drive a tortoise. Yet the Findorff crew remained intrigued, and they expressed interest in making the next wall map on their own.

The first two scaffolds of the next week wobbled, so quality and then speed suffered. The crew averaged five hundred bricks a day. On the other hand, they began behaving as if they'd figured the rhythm out and were on cruise control. The wobbles made Scott want to push new software (which featured Glenn's updated dynamic compensation), but on account of the risk, he restrained himself. Instead, Findorff agreed to ditch the short Hydro-Mobiles and stick with ninety-footers. That left only two walls.

Addressing the penultimate wall, a Findorff mason was blinded by glare on the HMI's screen and hit the wrong button. It was the last day of summer. The arm reached somewhere it shouldn't have and crashed into SAM's central table, its sternum. After that, the robot froze three times.

Autumn arrived and brought two days of rain.

The temperature dropped ten degrees.

When the cruise-control crew returned, it was as if they'd found a new gear. For two hours, they ran at 330bph—and then they were done.

Findorff reported being 20 percent under budget, but they also reported that without all the Hydro-Mobile maneuvers, they would have been 30 percent under. SAM, on so many short walls, had slowed the crew down. But Findorff also said the magic number was 40 percent. If SAM could raise productivity 40 percent, they said, they'd be thrilled. Eliminate the hiccups and headaches, they said, and SAM would be pretty close to viable. That wasn't the worst feedback from a first-time job.

Findorff also mentioned SAM in a couple of press releases, one of which showed not only SAM's bright red arm but also its bright yellow backup air compressor. The text was clear, almost insulting in its reliance on the future rather than the present tense. "Ultimately," it read, "this new technology will help lower health and safety impacts on the workforce, and provide consistent and reliable quality." *Will* help, not *helps*. Still, the enthusiasm was not feigned.

As one season passed to the next, Scott ceded that CR had learned a few things. But really, he'd only relearned what he already knew: that SAM didn't make sense on short walls. He knew it from Tennessee, and he'd known it five months before, when he and Nate had told Findorff's execs that SAM wouldn't work on every job. But this school was what Findorff had, and it had been impossible to turn down a company like Findorff.

SAM hadn't laid 1,700 bricks, as Scott had once promised, but it did lay 1,400 (in only four hours) and did so entirely under the operation of a foreign hand. Even in Missouri, the outside operator had undergone some serious tutelage. Not in Wisconsin.

Scott suspected he'd hear from Findorff again—not immediately, but soon.

22.

Desperation

That October, Scott and Zak scrounged as they hadn't before: simultaneously. Scott, with his tail between his legs, reapproached Soderberg Masonry—of the brutal Laramie job—in hopes of landing the one job that seemed destined to secure SAM's fame. It was in Boulder, Colorado, where Google was cladding a swanky new three-hundred-thousand-square-foot office building with intricate masonry. Soderberg had the contract, but if anyone recognized the potential of automation, surely it was Google. In Scott's fantasies, Google—which had an appetite for innovation and robotics—would see SAM run, know right away that it was looking at the future, and extend its mighty pedicured tentacles, propelling the start-up to the stratosphere.

Zak, meanwhile, flew halfway around the world to the United Arab Emirates, to represent Construction Robotics at an intriguing new program called the Dubai Future Accelerators. Flashy hotshot upstarts including Uber and Hyperloop One had been accepted—and just being mentioned alongside them made it especially tantalizing to Zak. Initially, Scott had been lukewarm about the idea; it didn't help that his wife did not want him huddled away in the Middle East for two months. Nate also had been opposed. He'd called the program a waste of time. He wanted jobs with American construction companies, not "support" from nebulous foreign organizations, even if they were royally funded. His company barely had a toehold in America! Who knew how they built buildings over there.

Even Zak had admitted doubts. Sometimes the program seemed

as cobbled together as the Alpha machine, no more substantial than its flashy website. Then an administrator had called and told Zak that anything was possible. He kept using that phrase. Dubai was booming, veritably oozing money, and SAM, he said, could be just the thing that builders there were looking for. Among other things, Zak learned that a building shaped like a huge picture frame was going up. "Maybe they pay us to modify SAM so it fits their bricks," Zak said, "then they buy two machines and four Hydros and build something crazy." Understandably, he wanted to be the guy who steered CR toward success; he couldn't help thinking of the bonus he'd earn if his lead metamorphosed into one or more sales. Maybe he'd use the money for a new truck of his own. Besides, there had been little to lose: All costs—airfare, hotel, office space—were covered.

Because nobody had been sure who would best represent CR in Dubai, everyone in the office had submitted his passport for a visa. Employees could swap out, the administrator said. Glenn had been hesitant. He'd read a story online about an Australian woman who had claimed she'd been raped, and then couldn't find four witnesses, and so was charged with having sex out of wedlock, and jailed. She'd been in jail for two years. This did not please him, and it didn't help that Dubai was regularly described as the Las Vegas of the Middle East. Still, he'd brought in his passport. Zak, for his part, recognized that life in Dubai would be constrained—no swearing, no porn—but even there you could grab a beer at the hotel bar. He just needed, as he put it, not to fuck up. He kept returning to the unconstrained potential. "I could be a hero!" he said. "It's like, the other day I was buying something, and didn't have any change, and a guy gave me fifteen cents. No big deal. Five million dollars to the prince is like the same thing. It's like pocket change."

Zak decided to take Kim, the gripper engineer, along with him. When they landed in Dubai, it was 110 degrees. They were given a two-bedroom suite on the ninth floor of a sleek glass tower called the Warwick Hotel. Zak, feeling noble, let Kim have the larger room. At a government building, they were also given a first-floor office. It was in a public area, so from the get-go, they received a lot of eyeballs. In such context, Zak wore a suit every day—partly because it was appropriate, and partly because Scott had specifically instructed him not to show up in shorts.

Because Zak and Kim didn't bring an actual three-thousand-pound bricklaying machine with them—much to Zak's chagrin—they had 3-D-printed a dozen models of SAM. Each was eight inches long, the size of the toy robots that Scott had played with as a kid. Nifty as they were, the models reduced an actual technology into a concept. This left Zak and Kim with posters and brochures and videos and some little toy models and all but made them ask people if they could simply imagine the rest. They might as well have brought *Goodnight, Goodnight, Construction Site*. From the outset, this made CR seem less advanced than it was—especially because CR's peers in the program included many entities with more vision than substance. One envisioned an array of modular electric cars like public trolleys, but slicker and with coffee bars. One made a new eco-friendly material that could be rendered into, it said, nearly anything. One company was developing a new research-funding platform, and one was developing new education standards. In such company, Zak and Kim seemed pie-in-the-sky idea men.

During the first week, program organizers paraded so many visitors into CR's Dubai office that Zak found it hard to focus. There was a new visitor every half hour, and the conversations were taking familiar contours:

PROGRAM ADMIN: *Come this way and see the bricklaying robot.*

ZAK: *Hi, I'm Zak, and this is SAM. Well, a model of SAM.*

VISITOR: *What are you doing with SAM in Dubai?*

ZAK: *Apparently just talking about it.*

Then, in mid-October, the program officially began. The program's organizers informed Zak that while two VIPs were escorted through the building, and shown all the impressive businesses that had been gathered in the up-and-coming UAE, he was not to open his mouth. This irritated Zak, because he knew that the VIPs were the crown prince and the prime minister. These were the men with serious pocket change. Nevertheless, Zak followed orders: He remained at CR's post until the entourage sauntered his way, surrounding him as they examined posters and brochures. The then crown prince—barely older than Zak—made eye contact, and Zak struck up a conversation. It was impossible not to. The prime minister, much older, looked at Zak, too. The crown prince asked some questions about SAM and then declared, "We need this technology." Zak may have once flubbed his interviews with mid-Atlantic masons, but he did not flub his brush with Middle Eastern royalty. His Supremacy the Prime Minister said nothing.

By the end of the first week, nothing was the big word. Precisely nothing was being built, or funded, or accelerated. In fact, the trend was in the opposite direction. Zak realized he'd been roped into an elaborate show. But because there were worse things than poodling, he tried to stay optimistic and patient. Amid the poodling, he struggled to stay on top of his usual work but found that his laptop was not on par with his array of monitors back home. Here and there, he wondered if, after all these pointless conversations, something might yet happen.

To take their minds off the situation, he and Kim decided to go to Ski Dubai. If they'd been looking for a better simulacrum of Las Vegas, they couldn't have found it. Ski Dubai was an indoor "ski area," a tilted warehouse full of man-made snow. The longest run was a quarter mile. The vertical drop was just under two hundred feet. One content with so little slope and gradient might as well ski the Brooklyn Bridge. The place had one chairlift. To Zak, who skied in Utah, it was pathetic, and he lowered his expectations accordingly, coughing up eighty dollars for a two-hour pass. It took an hour just to get the pass and gear. (A hat and gloves were not included. Not like it was Syracuse cold.) He put on a ridiculous blue jumpsuit, strapped on his boots, and stepped into the bindings. Amid the crowd, he shuffled toward the front of the lift line. He rode the lift up, then slid off—and very quickly was at the bottom of the slope. The slope—which would be a beginner run at any real ski area—was bestrewn with wild and injured skiers. It was comical. Zak was easily the best skier there. At the bottom was a pile of skis like Zak had never seen before. Everything about the place screamed shitshow. Zak did two more runs. They were not amazing. They took at most forty-five seconds each. His bare hands never got cold.

The pass also included a session of indoor skydiving (over a huge fan), so Zak and Kim did that, too, at a different facility. Different gear, different position, different brief thrill. These miniature excitements did not rival the one that had brought Zak and Kim to the Emirates. Playing with gravity was okay, but playing with the future was more thrilling. They wanted to place SAM on the starting line of a very long run.

Afterward, they found a store the size of Walmart and stocked up. Among other necessities, Zak bought an electric hot plate for twenty dollars. Back at the hotel, he emptied the minibar fridge and filled it up with cereal, milk, eggs, and meat, which he would eat for the rest of his stay. Though Zak found Middle Eastern food appealing, with all of its fried meats, he cooked himself a burger for dinner almost every night.

During the second week, Zak perked up as a field trip to construction sites was arranged, the perfect opportunity for him and Kim to assess SAM's utility in Dubai. The construction sites were like nothing Zak had ever seen before. They were crawling with men—nearly five thousand men. Workers stood around everywhere. Each scene was wilder than the ski hill. When Zak mentioned SAM, his tour guide asked why they would pay for a machine when they could just bring in more people. *Why did you pay to bring me here*, Zak wondered, *when you could have just told me this in an email?* At any rate, the fact before Zak was that laborers in Dubai earned the equivalent of only two dollars per hour—wages so low that they rendered SAM an untenable proposition. That and nobody was building with brick. There was no clay anywhere—just sand. Blocks yes, but bricks no. Also, Zak learned, because the government didn't build its own buildings, it would never buy SAM. A government rep suggested that he develop a flower-planting machine instead.

Late one night, Zak called Scott and told him what he'd seen. "There were forty-five hundred workers on each jobsite!" he said. "That's a fucking anthill," Scott replied, and asked about the building process. "The process," Zak said, "is a fuckton of lazy workers taking their sweet-ass time to lay block." This was not just the language of brothers-in-law but of experienced hard hats—although success had eluded them, they'd almost unknowingly become experienced.

The rest of the week passed in a blur of frustration and boredom. By the end of the third week, the office had cleared out. A third of the companies admitted to the program had bailed, and the rest had stopped participating in show-and-tell. Zak considered bailing but did not because he detected glimmers of hope. When the U.S. ambassador stopped by, the program head directed him to just two companies: Hyperloop One and Construction Robotics. To Zak, the head of the program kept saying that SAM was great, wonderful, the quintessence of the program. Privately, the head said that the crown prince's response was encouraging and that CR was on track to get funding. More specifically, he said that CR was on the shortlist for establishing a memorandum of understanding—though it was never clear what exactly that understanding would entail.

Days passed, mostly in front of a laptop. Zak answered emails from Scott and stitched together a proposal to develop a block-laying machine, since SAM was moot. He dreamed of building a factory in which robots prefabricated walls. Scott reminded him that building a factory in a foreign country was unrealistic—the most they could hope to do was modernize (make safer) jobsites with one complicated enough machine. At one point, Zak and Scott discussed offering His Royal Highness naming rights for this hypothetical block machine. Maybe they could call it Dublock.

One Sunday night, Zak headed to the hotel bar, got an expensive beer, and watched a football game. It was an afternoon game, but in Dubai it started at nine p.m. Everything, it seemed, was upside down. One night, he went golfing with the CEO of another tech company. Tee time was eight-thirty p.m. The eighteen-hole course was fully lit up. They played until midnight. He watched movies on the screen of a tiny laptop and conversed via FaceTime with the girlfriend he'd left behind in New York. Meanwhile, he missed his niece's fourth birthday.

One weekend, tired of the city, Zak and two other guys rented a car and drove twelve hours to the beach in Oman. In Muscat, they admired the Sultan Qaboos Grand Mosque, which had taken six years to build and whose masonry dome easily outflanked the best in America. They lay on towels on the beach and soaked up the sun while, not fifty feet away, one wave at a time, water ground particles into unusable construction material.

At some point, back in Dubai, a housekeeper found eggshells in

Zak's trash can and informed security. Zak was told not to use the electric hot plate in his room anymore. He ignored the instruction and continued to cook his own meals until the last day of the program, when hotel staff confiscated the device. By then, Zak had eaten at least thirty burgers, and come no closer to getting SAM or CR some action.

When Zak returned home for Thanksgiving, he rediscovered the sensation called cold. He wasn't used to it anymore. His first words to me—back in the land of permissible swearing—were "Fuck them." Everything, he said, was last-minute and disrespectful—and all for nothing. "We got shortchanged. They're like, 'We're so innovative,' but the whole program was fucking ridiculous." He was furious that CR had been paired with Dubai Municipality, an entity that didn't know a Hydro-Mobile from a hydraulic jack. All it did was analyze the construction industry. His voice rose. "They just—they, they, they—" He got so mad he began stuttering. "We asked for simple stuff and never got it. We wasted a lot of time, and I'm not sure if anything will come out of it. Which is what I was afraid of."

Worse, Zak felt that Dubai Municipality was a joke. Where it could have funded any of two dozen innovative companies—companies with actual products that were brought a great distance so that their cases could be made—it funded a company called Grow, which intended to cultivate coffee hydroponically. Where was the IP in that? The idea was no more compelling or novel than opening a new sushi restaurant in Sarasota. The whole thing seemed to be a sham.

More evidence: Those promised MOUs, he'd heard, led nowhere. They were fluff. All they said was: *We agree to work with you for the next sixty days.* And the big money—the royal pocket change—was never there. Hyperloop One, as Zak once believed, didn't actually get $50 million from the accelerator. It got the money independently, before the program ever started, but the press releases didn't exactly clarify that. All in all, he felt duped. "If we hadn't gone, we'd have been like, *Oh my God, we missed out on such an opportunity*," he said. "It'd be like knowing the winning lottery numbers and not buying tickets." Now he was just pissed.

In Fort Collins, Scott was trying to play things tactically. He was sitting at a rectangular oak table in a small spartan room in the offices of Soderberg Masonry, across from the company's owners, John and Chuck Nacos, and two of their estimators. The vibe was far from Findorff or even Brawdy. There was barely enough room to pull the chairs out. On the wall behind the Soderberg men was a poster of a chisel. The tool was as long as a human leg, and beaten to hell, and set against a background the color of wine. In modernist fashion, text across the bottom of the poster said LIGHT GRAY CHISEL. When, after handshakes, Scott passed out brochures describing SAM, he thought it was almost too bad they didn't feature the text COMPLICATED BRICKLAYING MACHINE.

Scott took out his laptop and handed out business cards, then asked to keep the last one, since he'd run out. There was a small window behind his left shoulder, and through it, from across the parking lot, eau de coffee roaster wafted into the room. It smelled less like coffee beans than burned toast. Audibly, a wall clock tick-tick-ticked.

Scott had come west for a friend's wedding, but he'd allocated a day to pursue some leads, because the region was booming. Nearly everywhere you looked along Colorado's Front Range, projects rose from the earth. From almost anywhere in Denver, a glance at the skyline revealed no fewer than ten tower cranes. And Denver was a brick town. There was so much construction that it made Scott want to establish a western hub. From there, CR would have access to Salt Lake City, even Seattle. That morning, he'd met Denver's Hydro-Mobile dealer and been advised on which local builders to pursue. The dealer had pointed him toward Mortenson, one of the biggest builders in the state, and he'd dropped by. But the dealer had not encouraged Scott to visit Soderberg, because they generally used pipe scaffolding, even on buildings hundreds of feet high. But Scott didn't want to write them off. Soderberg was one of the top three masonry companies in the tristate area (Colorado, Wyoming, Utah), and notwithstanding its aversion to modern scaffolds, the company had rented some sixteen months before to give SAM a try. That said a lot. So did John Nacos's words, delivered by phone only a month after the job, that he was willing to try SAM again.

Scott knew exactly where he wanted that next trial to take place, and he was thinking of another way to make the case. The previous June, Soderberg had agreed to pay CR a dollar a brick. The company had put down ten thousand dollars for ten thousand bricks, and even though SAM laid only sixty-six hundred, Soderberg never asked for the difference back. Scott thought he might offer Soderberg that credit toward the Google job. He'd never actually met John and Chuck Nacos, but it seemed a good sign that they'd offered him forty-five minutes to make his case. Scott was pretty sure that they wouldn't have invited him in just to say no or laugh in his face.

Before the meeting, Scott had looked at the data from the Laramie job. On SAM's best day, the machine had placed 932 bricks in six hours. "If we did that now," he said, "it'd be eighteen hundred. We're so much faster now." This got him going. "It's one of those jobs you wish you had over again. Well, every job I want over again." But before he invented a time machine, what he really wanted was Soderberg's Google job.

"It's been a big year for us," he began. "When we worked with you, we were laying about two hundred bricks per hour. Now we're closer to the range between three-fifty and four hundred."

One of the estimators interjected, "When she's running good."

"Yeah," Scott said, ignoring the gender assignment, "but we're even getting three-twenty on walls with windows. We're focused on productivity, ease of use, setup, and fixing bugs. I'm not saying problems don't happen—it's construction . . . But now we have a training program." This program, he said, meant "you come to Victor, learn setup, which is lots easier than it was in Laramie, how to make maps, run the machine." It was very easy, Scott said, and took only three or four days.

He cited evidence. "We were on a job in Wisconsin this summer; my guy was there but didn't run the machine at all. . . . On a job now, my guy is just there for support." He continued, "The track is easier to install. There's no welding. You can preattach it in your yard, ship it to the site, then fold it down and tighten turnbuckles on-site. We are definitely head and shoulders above where we were a year ago." Calibration, measuring, and consistency were all improved, he said. And: "We appreciated the opportunity to work with you and would love the opportunity to work with you again and show you what we can do."

Scott leaned back in his chair. His laptop was still closed. Across the table, John, in a black short-sleeve shirt, had his arms crossed. Forty years old, he'd been a bricklayer. Chuck, John's father, had been, too, before he bought Soderberg from a Swede. Behind wire-rimmed glasses, he was not making eye contact; he was just taking notes, looking down in thought.

"A guy had to follow behind and tap 'em all," the estimator said. "Is it better?"

Significantly, Scott said, "You should not have to tap one hundred percent. If you're tapping more than ten percent, you should calibrate the machine . . . We took a whole month and just did a deep dive on quality."

With this, Chuck turned to John and asked about upcoming jobs. Chuck pondered out loud: "Not Platte, or Foothills, or Civica. Not Welton. York is too cut up. Woulda done it at Metro State."

Scott asked: "Google?"

"I thought about that," John said, "but it's not set up for HydroMobiles. I thought we could all get famous."

Scott remained poised, and then the estimator suggested a high-rise apartment building in Denver, at 375 South Jackson Street. Collectively, eyes on the Soderberg side of the table opened wide in excitement. A roll of architectural drawings was brought in and unfurled. Everyone stood up and leaned forward.

"We're bidding on this project tomorrow," John said. The plans showed a ten-story building, eighty-two feet wide, clad in large bricks. Scott flipped pages while chawing on a wad of gum.

"Holy windows!" John said. "We have to build this thing? Jesus!"

"That's gonna kill your productivity," Scott said. He added that the alternating bond pattern was a no-brainer for SAM. "I could run some numbers," he offered, and John said he'd send the plans over.

"It might be fun," the first estimator said. He pointed to the wide facade and said there'd be lots of room for SAM to run.

"With the right attitude, the right training, and the right preparation, you can . . ." Scott trailed off and let the Soderberg guys come up with the word. The word was "profit." Or "dominate."

It was a step in the right direction. Still, doubts remained, and it was obvious. At one point, Chuck Nacos referred to SAM as "the damn thing." Scott let the insult slide. One of the estimators admitted that he thought SAM would never work and, on his iPad, pulled up a video he'd taken of SAM running in Laramie. Not even a year and a half old, it seemed paleolithic, on account of the laggard pace and antiquated buttering path. This Scott did not let slide. On his own laptop, he pulled up a comparison video that Zak had created. "Last year it was fifteen seconds," he said. "We're under nine seconds now." He mentioned SAM's improved software. Then, coyly, he tried one last time.

"Any chance the Google job is viable?"

"It'd be a lot nicer," John said, "if the plan was to incorporate Hydros."

Scott pushed on. "What is your estimated production?"

Soderberg was planning only three hundred bricks per day per mason. Hearing this, Scott nearly jumped. "It wouldn't take that many bricks per hour to make it profitable," he said.

"I like the idea," John said. "I talked to Chris Raddell a bunch about it. He's got plans."

Scott knew this but remained calm. He was there because, having submitted a proposal based on those plans, he'd received only radio silence. He said, "Would you be interested in an analysis?"

And then John backed off. "Lemme see where I'm at on it," he said. "It might be so far along. It's coming up in December." Then he turned the question on Scott: How come he hadn't built something big and elaborate? Something like the Google building?

Scott said, "Limited resources, you know? I'm sure we'd kill it, but we need to partner, because we don't self-perform. We just build the technology." He steered from the abstract Google building to the real one. "If we go through an analysis, we'll make sure the numbers work."

"Google. Boulder," John said. "It works." Did he know that Google had acquired $4 billion worth of robotics companies the year before? It more than worked.

Then, although Scott had thrown pretty much every possible argument at Soderberg—having decreed that CR had gained wisdom from experience, that Hydro-Mobile ownership was not mandatory, and that in a strong market, SAM was a no-brainer—he threw more. "We've come a long way with the ability to set up, move, and get running quickly. I think you'll be impressed with how far it's come. If we get the opportunity to work with you again, I'll make sure it works." He didn't offer the $3,400 credit.

"We're right with you," John said.

"I'd like to see it again," the first estimator said. John couldn't say that, because in June 2015, he never actually saw SAM run.

With time running out, Scott all but begged for the job in Boulder. "I'll be honest," he said, pointing to the Jackson Street drawings. "There are parts I wouldn't advise you to use the robot on. Not for your first job." Scott immediately wished he could take back the words, because the guys across the table clearly remembered their first job with SAM.

And then the clock struck four p.m. They got up and shuffled toward the door. On the way out, in the hallway, Scott noticed a poster of Coors Field, signed by Rockies coach Don Baylor. Soderberg had built Coors Field. John Nacos said the brick pattern was so tricky that it took their mason a while to figure out how to build it. Scott laughed, because he knew the pattern. Zak had found it online, and Scott had liked it so much that he'd had Glenn teach SAM how to make it. Only later did he realize Soderberg's involvement.

They shook hands in the hallway. Scott walked out, got in his rental car, and backed out of his parking space. He put it in drive. He was traveling again. As he passed the entrance gate, he noticed a huge stack of pipe scaffolding, which made him grimace.

His phone rang. It was Zak, in Dubai.

"How'd the meeting go?" Zak asked.

"I think we missed our opportunity with the Google building," Scott said.

23.

Mississippi

Meanwhile, CR's most promising job yet was falling apart. The job was in Vicksburg, Mississippi, where Yates Construction was building a massive new headquarters for the Army Corps of Engineers. The whole job incorporated three hundred thousand bricks. It was SAM's biggest job to date—ten times bigger than the one in D.C.—with its biggest builder. Yates wasn't big so much as gargantuan. It was one of the two hundred largest companies in the country, the builder of industrial plants and ports, casinos and resorts. With revenues of $2.4 billion, it employed six thousand people, including numerous masonry crews. The company's research division had been watching SAM since Fort Lee, ready to pounce on the latest reliable innovation. Yates was finally embarking on a test run, which said a lot. If they liked SAM, they could very well buy dozens of machines and outfit the masonry division over years.

Better still, the job was being run by a young mason who believed as firmly in SAM as Scott. The alignment of their vision was unprecedented, really. In most of a decade, Scott had not met anyone as enthusiastic about SAM as David Hux.

They'd met at the Tennessee job, where David and his boss had come to scrutinize SAM. David's boss, a minister, was not a sudden convert, but David argued that it was unfair to judge SAM on someone else's performance. Scott, standing beside them on the scaffold, heard this and thought, *YES. A true believer*. Scott invited him to Victor for a test drive, as it were. David, who lived thirty miles from the nearest traffic light, had never been on a plane. To see the future, he boarded a Boeing and flew to New York.

In New York, Scott and David became so fond of each other that Scott's wife quickly grew irritated. "Who's your new best friend from Mississippi?" she'd ask in a voice tinged with jealousy. His new best friend was the same age but short, hefty, and encumbered by a drawl. The youngest of eight brothers, David was tough, having worked his whole life on construction sites and oil rigs. His apprenticeship had started early. As an eleven-year-old, he'd worked as a laborer. He was too short to lift a wheelbarrow, so the crew had him haul mud and bricks in a little red wagon. By thirteen, he was a real laborer. By fifteen, he was laying bricks. He didn't finish high school and didn't go to college, but he could lay a thousand a day. His personal record was eighteen hundred. He named his trowels Excalibur but went through a number of them. Each would start out thirteen inches long, and when he scraped it down to nine inches, it was time to get a new one. The process would take him nine months. He had another way to prove his toughness: his ability to drink. He'd lay a thousand, drink a case of beer, and do this every night, all week. He became a 350-pounder, bigger than the Destroyer.

He started a masonry company and offered his masons a nickel for every brick beyond 480. Motivation like that got his men to 1,500 bricks per day. That was extra pocket cash of $250 every week. He didn't know it, but it was the same bonus arrangement Garrett Hood—the winningest Bricklayer 500 champion—employed in North Carolina.

David Hux knew he could outlay SAM, but he didn't want to. He'd undergone two surgeries on his shoulders, as well as a few others. He was a realist and knew the industry desperately needed to attract young talent. All of the masons he'd seen down south were old, their days of hauling eighty-pound bags of mud and laying bricks in 105-degree heat limited. SAM captivated him. He thought he was SAM's biggest fan, and he thought that SAM's potential had barely been tapped into. He was gung ho about the machine. Where Scott had grand dreams, David's were even grander: He could imagine finishing a fifty-thousand-brick job in two weeks by throwing two alternating crews at it. In every eight-hour shift, each crew would lay three thousand bricks, sometimes thirty-five hundred. He might have been a country boy, but he was no technophobic hick. He thought SAM could spark a revolution in the construction industry. Nearly note for note, he had the same vision Scott had been chasing for nine years. As he put it, "We seen thangs ahh to ahh."

Scott's encounter with David Hux seemed all the more serendipitous when, one evening, they ended up scheming at a small sports bar called Smokin' Joe's. There, while Scott drank bourbon and David drank SoCo Lights (half Southern Comfort, half Diet Coke), the summer Olympics unfolded on a TV not five feet from their table. On that iridescent screen, it had been impossible not to notice a remarkable performance by the American Katie Ledecky. She was swimming the 800-meter freestyle, a race Scott knew all too well.

Ledecky's arms and shoulders were as big as a mason's, but she swam like a machine, with an uneven, loping stroke. Her speed was incredible: She swam so fast that NBC couldn't handle it. By 400 meters, in order to keep Ledecky in the same shot as the field she'd been leaving behind from the first stroke, the cameramen were forced to zoom out. By 750 meters, they couldn't zoom out far enough. To give viewers a shot of the final turn, NBC abandoned Ledecky entirely. Churning, she swam straight off the screen in one direction, while the field swam the other way. She finished half a length ahead of the world's fastest swimmers, eleven seconds ahead of the nearest competition, and set a new world record of 8:04.79. As Ledecky ripped off her cap and raised her goggles, the camera stayed on her, because she'd turned an event typically thought of as a slog into a thrill. She leaned on a lane rope and took twenty big breaths, then was all smiles. Meanwhile, commentators struggled to put her performance in perspective. They called Ledecky blistering, a perfectionist, a superwoman "without peers, now or ever before." They compared her to Secretariat, a hydrofoil, and a torpedo. But the comparisons weren't apt, and Scott knew it.

Ledecky remade the field of distance swimming by swimming like a metronome. She'd studied her stroke length and found that she swam her fastest if she held it at 1.36 seconds. So just then, for half a mile, she'd pegged it there, plus or minus a few hundredths of a second. At that cadence, she'd traversed each length of the pool—each 50 meters—in 20 or 20.5 strokes. No more, no less. Left, breathe, right . . . left, breathe, right . . . left, breathe, right . . . length after length. Her stroke had been the fastest in the pool, and she'd maintained her stroke count sixteen times in a row, without slowing. She trained at that pace, always swam at it. She ticked off each 100 meters in 1:01, as if not tiring, even turned up the speed on the final length. In this fashion, Ledecky won elite trials of speed by margins of 1 or 2 percent, earning her the title of the most dominant athlete in sport—more than Bjørn Dæhlie dominated cross-country skiing, more than Usain Bolt dominated sprinting, more than Lance Armstrong, who doped, ever dominated the Tour de France. The year before, Ledecky had swept every individual freestyle event (200m, 400m, 800m, 1500m) at the World Championships. There, in Rio, she won gold medals in the 200m, the 400m, and the 800m (setting world records in the last two), and won two more medals in relays. In one week, she became the most decorated woman at the 2016 games and the most decorated U.S. woman at *any* games.

The head of USA Swimming called our time the Ledecky era. He said we'd look back and the sport's history would be defined as Before Ledecky and After Ledecky.

Scott knew exactly how to describe the performance of the smart young East Coaster. She swam like SAM.

At that humble Rochester bar, all of it—the alcohol, the long-awaited enthusiasm, the inspirational example—collided, and gave Scott hope that finally, with David Hux and Yates, he'd found the path to success and recognition.

The job had begun well. On the first day, addressing a wall 140 feet wide and 65 feet tall, David Hux compelled SAM to lay 1,560 bricks. On the second day, not long after sunrise, he called his crew over to the middle of the scaffold and delivered a pep talk. He said he aimed to be a leader in working with the technology before him, because it was destiny. He called SAM the future. A religious man, he encouraged his crew to work together for a purpose. He made sure to find masons who didn't reveal a certain resistant attitude around SAM, and he helped them modify the way they worked. "All you do is make sure there's brick and mortar in SAM," he told a laborer. "If there's other shit to do, we'll have somebody else do it." He poured energy into running SAM. He directed so much energy toward SAM, in fact, that he forgot to order mud and bricks and propane. That day, the machine laid nearly a thousand bricks and left Stephen Kean doing little more than watching.

The third day was a bust, because the building's lintels stuck out so far that they blocked SAM's laser. But the fourth day was huge—2,770 bricks, which was a record by a significant margin. David Hux showed all the signs of having the magic touch, and it seemed very likely that when, in two days, Yates's president, William Yates himself, showed up in a helicopter, he would recognize as much.

Then, on the fifth day, the machine tore itself apart. It happened at a quarter after ten. SAM, at the time, was forty feet up, between two windows, in a long meaty stretch of wall. Under David Hux's direction, it was laying the third course of the day. SAM had been running under an hour and was about to lay the three hundredth brick. The Stäubli twisted, reached up for a half-brick in chute one, and jammed on a protruding nub of clay. Stephen stepped in, directed the Stäubli to its home position—and that was when it happened. As if taking a big stretch, the arm extended way up, and the cable dangling from it snagged and ripped out. The gripper went dead.

"Fuck," Stephen said. He tried to recover. The Stäubli responded, but the gripper did not. Its little green lights had gone out. The camera couldn't see, and the gripper's fingers would neither open nor close. Stephen realized very quickly that a Bad Thing had happened. This warranted informing Scott, so just before noon, he called his boss and provided the vaguest overview. Scott, at the time, was in Colorado, en route to Soderberg.

Stephen examined the situation more closely. The wires had not broken; they had just been ripped out of their connections. Stephen was relieved, assuming he could put the wires back if he could determine where each one went. He called Kerry. For forty minutes, Kerry walked Stephen through each wire's position. Stephen then rebooted

SAM and tried manipulating its business end. When the gripper did not respond, the severity of the situation sank in.

On the Hydro-Mobile, the men were starting to panic. Mostly black, they had a quick drawl that Stephen rarely was able to understand, but this he understood the first time: *How's it looking? Will the robot be ready soon?* Stephen wasn't sure. For every question his answer was: *I should know more in the next hour.* Experience had taught him to jog SAM off to the side, so that the masons could at the least tool the joints in the wall without reaching around a hippopotamus and keep laying bricks by hand. When the project manager asked if SAM would be ready for the president's visit the next day, Stephen told him the truth: *You might want to reschedule.* By then it had sunk in that the situation was beyond Bad—it was Very Bad.

Stephen called Kerry again, and they talked for another forty minutes. While talking, Stephen investigated, broaching new territory. He was poking inside Stäubli's arm—forbidden, proprietary, warranty-voiding territory. He had no choice. It was scorching up on the Hydro-Mobile, and Stephen, sunscreenless, was trying to stay cool and focus. Below him, trucks rumbled around the site, spewing water to keep the dust down. He burned through a Gatorade.

He discovered that the connectors—the male ends from the gripper and female ends from SAM—had somehow been damaged. He called Kerry six more times. It took Stephen two hours to figure out that although the wires fit together, the connectors had been bent. As a result, they jiggled, which meant the signals traveling through them flickered like a lightbulb.

What's the plan? the masons asked. Stephen's answer became: *I should know tomorrow.*

At three p.m., Stephen called Stäubli's tech support number and talked briefly while tinkering. Unfortunately, this Stäubli rep wasn't especially knowledgeable. He suggested a new robot harness, which cost $4,500. Stephen called Dave Kolczynski, the director of engineering, as well as Scott and Glenn. Thrice more he called Stäubli's tech support line, and since shifts had changed, he got Stäubli's other rep. Stephen liked this other guy, because he thought more mechanically. This rep advised Stephen on how to disassemble the arm, which was not exactly kosher. But digging around proved a fool's errand—Stephen discovered no shorts or breaks and could not reach the connector in SAM's forearm that he suspected was twisted.

He called Glenn, and Chris Johnson, and Kerry twice more. A plan emerged. Kerry would open up the arm on SAM6, remove its cable, and overnight the thing twelve hundred miles from Victor to Vicksburg. Extracting this cable and all of its component wires took Kerry an hour. He dropped the unit off at the UPS office in Victor at five-thirty and went home. After the Tennessee job, it all resembled déjà vu.

In the Bayou, the afternoon slipped away, and the construction site cleared out. All went quiet. A dozen telehandlers settled down, as did three times as many basket lifts. The mobile crane stood stationary. All of the various framers, welders, steelworkers, HVAC installers—even the dedicated safety crew—went home. Good night went the construction site. Technically, Stephen was no longer supposed to be there, not without a superintendent present, but he wasn't about to leave the robot in pieces. By then he'd removed the whole hand. He was trying to button things up without bungling things up. Twice he nearly dropped a screw. It wouldn't have been the first time. In Missouri, he climbed thirty feet down to look in the dirt for a driver bit he dropped—and found it. In Tennessee, he did not. At one point, his heart sank when he thought he'd lost a screw somewhere inside the Stäubli. It had been a long day. He was fine.

From the scaffold, he watched the sun set. All day, he'd consumed only six ounces of tuna fish and hot sauce, and three Gatorades. He didn't get dinner until ten p.m.

What grated on Stephen was that CR had put effort toward avoiding this very scenario. Only a couple months before, when they'd changed the path of the arm to speed it up, the engineers had discovered that the cable dangled oh so differently. It flopped a few more degrees, sagged, and wiggled a smidge closer to other parts of SAM. The first time it had snagged, Mike Oklevitch was testing the machine. The arm had been reaching for a utility brick in chute two, and the cable wrapped around the front corner of the chute. It was yanked clear out. It happened again. The guys had added a plate—a guard—to prevent the cable hang-up. They'd tested their fix, and even Mike

Oklevitch was happy with it. Apparently, they hadn't tested it enough, because that morning, the cable had snagged on that very guard. The irony hurt, and the situation bolstered Mike Oklevitch's belief that karma played a role in the whole endeavor: He had said that you make up for every stellar day the day after.

At eight the next morning, UPS delivered the cable to the Best Western. Things got lost when they were shipped to construction sites, so the hotel was the safe move. Dave, the head of engineering, texted Stephen at nine-thirty:

DK—*Did the new cable solve the problem?*
SK—*Still working on it. I should know in maybe 20 minutes.*

To an engineer, twenty minutes tended to be a bit longer. Besides, the task before Stephen was his most involved and technical with SAM. He compared it to an automobile's cylinder head. Installing the new wires, Stephen took care not to lose any parts or break anything. He was worried about causing more harm than good.

Three hours later, he responded to the text:

SK—*It looks like it's good now. Otherwise I'm going through the chute picks to see what I can do to fix the issue with the cable catching.*

This last act of prudence was nearly insubordinate. Everyone in Victor wanted to get up and running ASAP, so that Mr. Yates, even if he didn't actually see SAM, would at least see big numbers. Stephen resisted. He wanted to fix the problem so they didn't end up ripping out the new cable, too. Rushing into things, Stephen knew, had gotten CR in trouble before. He scrounged an aluminum flashing mount from the site, bent it, drilled holes in it, and connected it to SAM's arm using screws he scavenged from the smart pole. To this little bracket, he strapped the cable, holding it off to the side. Then he ran dry bricks for the rest of the day.

All of the hustling proved pointless. The president's trip was postponed indefinitely. After that, the wheels came off, leaving everyone else at Yates unimpressed. David Hux never kept SAM running for longer than four and a half hours, only once ran the machine in excess of two hundred bricks per hour, and only twice compelled it to lay more than a thousand bricks in a day. Where Yates's plan had been to use SAM for a month, and longer if things were looking up, the plan changed on day twelve. At that point, SAM's total brick count was barely eleven thousand. Yates called it, and away went more potential than Scott cared to contemplate.

24.

Indianapolis

By November, CR was focused on a new company out of Indiana: F. A. Wilhelm Construction. Wilhelm, an $800 million general contractor, wasn't quite Yates size, but it was bigger than Findorff and performed its own masonry work. By all counts, Wilhelm had the right guys with the right attitude and the right experience. In Victor, those guys had been trained to run SAM. In Indianapolis, they had a sweet job, bricking up the new training center for the Indiana Pacers. Wilhelm had been so eager that, in addition to signing up for three months with SAM, they manufactured their own SAM-friendly track for Pro-Series scaffolds. They made the stuff in their own shop for a fraction of what CR could have. Thanks to that, the job proceeded on schedule.

Unfortunately, the bricks were the worst Ken had seen. The scaffold was slanted. SAM's Wi-Fi acted up, apparently as frazzled by trains as by boats. The machine kept placing air bricks. The compressor sputtered. The laser ran away. The arm froze. The nozzle clogged, and buttered the first few courses so badly that they had to be torn down and re-laid. But these glitches paled to the latest: Where the machine had lagged fifteen seconds in Tennessee and Wisconsin, now it lagged sixty seconds, making SAM slower than the Alpha. It was as if the beast had lead poisoning. The crew, who had been so excited to get their hands on Construction Robotics T-shirts, very quickly became indifferent to the bricklaying robot.

Scott demanded a solution. While Glenn hunted for the problem,

Ken made smaller wall maps, hoping each would demand less processing power. This lasted all of one morning, until a component inside SAM's arm broke. The component was an amplifier, and without it, the Stäubli could run only at half speed. It was the same injury the machine had suffered a year before in Buffalo. Now SAM had torn a ligament and was left with a limp. This, on top of the neural delay, was too much for Scott, so a new plan was born. Because he knew better than to perform arm surgery in public again, SAM2 would be swapped with SAM4, which was parked in Mississippi. Where the early termination of the Yates job had left Scott disappointed that he didn't have two machines running well at once, the one upside was the machine's availability as a backup.

Ken struggled through two more miserable days, deemed an intolerable shitshow by Scott, before Stephen showed up, machine in tow. By the time he was ready to swap machines, the crew thought they had a pile of junk on their hands. Frustration levels soared. The problems, Scott admitted, were a pain in the ass. But swapping machines presented its own public relations risk. CR would look foolish if it swapped one craptastic machine for another. That would be a Halloween trick that nobody would appreciate. Scott insisted that Ken thoroughly check SAM4 before putting it on the scaffold. Triple-check everything, Scott said. Quadruple-check.

All the while, Scott told Ken over the phone to encourage the crew, take control, and stay positive even as the opinion that SAM was a piece of crap began to congeal. He knew that everyone was frustrated, but he also knew that his scrappy little start-up could provide software and hardware fixes so quickly that the pain would soon be forgotten. It was as close to a pep talk as Scott ever got.

Indianapolis wasn't Laramie, though—there was no room to spare on the site. Ken brought SAM4 to the Comfort Inn and, in the parking lot, fixed the air hose that had been damaged in unloading, and gave the machine a full physical. He went through the preflight checklist. He even pulled some unrusty parts off SAM2 and swapped them with the rusty equivalents on SAM4. When he brought SAM4 to the jobsite, he ran some quick morning air bricks to soothe Scott's nerves.

And then everything improved. SAM laid sixteen hundred bricks and ran, in Ken's opinion, 90 percent better than it had a week before. Scott showed up, ready to play diplomat, but sensed no frustration among the Wilhelm crew. He liked the attitude he saw, and they liked anything over fifteen hundred bricks. That day, he said, was the good day they needed.

Ken drove the broken machine back to Victor, and Stephen showed up in Indianapolis to fill in. Scott hung around, switching between playing the nervous foreman and the micromanaging CEO. Desperate for the job to go well, he paid attention to every detail: the raising of the scaffold, the moving of the poles, the making of the wall maps.

There was a tense moment when Stephen reported to Glenn that SAM4 had frozen half a dozen times—but the machine never crapped out, and it delivered thirteen hundred bricks. Wilhelm's crew insisted on running it themselves, so he handed over the tablet. They did well. By the end of the week, Glenn had found the root of the lagging bug and fixed it. The engineering team even put an app on the tablet for super-fast reconnection to the Wi-Fi. Remotely, they updated the tablet's driver, and at Office Depot, Scott bought a new Wi-Fi card for it.

And then, four weeks into the job, the Wilhelm crew attacked a new wall. It was the longest setup Stephen had seen, what everyone deemed a money wall. On Monday, Wilhelm's crew laid 2,000 bricks in six hours sans issues. The day went exactly as Scott had hoped it would. He was happy to be sitting around, happy that Stephen was twiddling his thumbs, too—if that was what support looked like, so be it. Sitting around would show that his bricklaying machine was worthy. The next day, Wilhelm's crew laid another 2,000, and pulled the job ahead of schedule. On Wednesday, 1,800. On Thursday, 1,500.

It was almost as if the machine, by its own intent, had been waiting for Election Day to see how Americans felt about change. Once the votes were in, the machine set to work, buoyed by support. Maybe the machine found more inspiration from basketball, whose evolution was that much more dynamic than that of baseball or swimming. Or maybe it was Nate's presence, his first visit to a jobsite in a year. Nearly seventy, with a back that ached more than ever, he climbed up and down the scaffold the only way possible, making Zak nervous every time. But he couldn't help himself. His faith was proving warranted.

The next day, less than a week before Thanksgiving, was unusually warm and splendid, a far cry from what nature had provided during the PMD job three years earlier. SAM was humming along, and the Peters-Podkaminer clan was in attendance. Zak, up on the scaffold, noticed that SAM was about to run over a mason's trowel, so—aware of the risk of setting his paws on a union bricklayer's possession—he grabbed it, held it in front of him with arms outstretched, as if a supplicant, and said, "Here's your trowel, sir." A blowup was averted. A trowel was saved. Stephen foresaw a day of "clear sailing," and that was what they got.

Wilhelm's crew compelled SAM to place 2,945 bricks, a new record by 200. The count prompted high fives, which do not occur often at day's end on jobsites. The crew, though, had outlaid twice as many men on the other side of the site, who were hustling not to get beaten by a machine. Mike Berrisford, Wilhelm's masonry operations manager, took note of his double win, just as one of PMD's owners had years before. And just as Scott had predicted, Mike Berrisford forgot the weeks behind him as something clicked. He thought, *I gotta get SAM, because look what it does! It invigorates my guys, and I get huge results.*

That same day, John Nolan stopped by the office in Victor and dropped off a case of Genny. He ran up the old steep stairs, hoping to find Scott and Tim and Glenn and Kim and even Zak, but they were all away. The oldest colleagues around were Dave and Chris Johnson and an intern, so John sat down beside them. He learned of improvements and innovations and further speed, then told the guys about all of the development and construction under way at the brewery, and what it was like to turn water into beer, as men had been doing for millennia. Dave opened one immediately and raised it as if it were champagne.

Scott returned to Rochester, and as another construction season wound down, he arranged two straight weeks at home. No nine-hour drives, no early-morning flights. All the while, he kept tabs on Wilhelm. As Wilhelm built an eight-day wall in five days, Scott began building an ice rink in his backyard, as he'd always wanted to do. His mother-in-law had given him a kit, with posts and a big plastic liner; with Zak's help, he bought a bunch of wood and got to corralling and controlling a pool of water. The key to making good ice, it turned out, was adding just a little bit of water at a time. After the last twenty-four months, Scott understood incremental development just fine—and was pleased to see something good come from weather conditions that otherwise would have frustrated him.

Less than a week into December, Wilhelm said it wanted to buy SAM.

At long last, Scott got what he'd always wanted: a mason contractor who'd solved the riddle of keeping SAM running, busy, and profitable.

Scott immediately wrote an email to everyone in the company: "Today we entered a new chapter. Today we earned a sale to Wilhelm because the machine that we all have worked so hard to create is proving it's [*sic*] worth." Congratulations and praise, he wrote, went to everyone. He called the future bright. He did not mention that the magic number of bricks in the wall was almost exactly what he had been dreaming of for a decade. This was, in the history of Construction Robotics, the most gracious and selfless act of negligence ever. Because he couldn't help it, he rambled on about improving SAM further before he thanked everyone. It was pure Scott.

Nate soon replied-all and wrote that it was as if CR were metaphorically atop Maine's Cadillac Mountain—a dome of durable granite in the tumultuous North Atlantic—catching the country's first resplendent rays of sun.

Epilogue

With that, the era of the MVP was officially over.

But by the time Wilhelm sent a purchase order to Construction Robotics, the little start-up's finances were, in the phrasing of the book that had guided the firm's long endeavor, minimally viable. CR's runway was down to just a few months. To Scott, his company's existence felt ethereal. For months, Zak had been putting off paying whatever bills he could, and even considered selling the big gray Stäubli from the Alpha machine, which for four years had done little more than collect dust. At one point, Scott worried that he'd have to move out of PMD's leaky office and back into a trailer in the parking lot. It was a great relief when a second round of NSF funding came through, but it was not enough. It was only on account of Nate that the company, in limbo for so long, had remained solvent. Monthly, he was writing six-figure checks to CR, keeping it afloat. As ever, he remained full of faith.

Scott's faith, for what it was worth, had wavered as he realized that the construction industry was far more conservative than he'd imagined. For years, he'd found it nearly impossible to find innovative firms. He'd looked everywhere. He'd asked brick distributors. He'd asked Hydro-Mobile dealers. He'd even attended a three-day conference focused solely on innovation in construction, where a famous keynote speaker promoted "moon-shot thinking" and asked: "How long before there's a robot on your jobsites?" Apparently, the keynote speaker didn't know that the tall man with the buzz cut, standing not fifty feet away, had already been bringing a robot to jobsites for a year and a half. So much for innovation in construction. In the audience were half of the men who'd built the world's tallest buildings, and even they could not steer Scott to the firm of his dreams.

Meanwhile, an upstart called Uber surpassed the value of General Motors, and disruptors in every other industry became all the rage.

For a while, CR had sought out the big boys, for whom a five-hundred-thousand-dollar machine was presumably peanuts, and then it sought out little firms, who presumably sought a competitive advantage. All the while, predicting their reactions remained impossible. One day, a guy would spit in Zak's face: *My guys lay two thousand bricks all the time! Who needs a robot for that?* Another day, Zak or Scott or Nate or Chris Raddell would stumble upon what seemed to be the best company ever. Any one of them would fall in love in an instant and wonder why he hadn't heard of the firm before and if there were others in the same mold elsewhere. This thinking only exasperated Scott. He'd wonder why firms in Rochester, which laid claim to so much innovation, were so averse to experimenting with modern technology, while those in D.C., a city predicated on awareness of the outside world, seemed aloof; and why, no matter where firms were, it was impossible to get anything moving via phone or email. The industry was plain old-fashioned. Convincing masons to buy SAM was, as Scott put it after a couple years of trying, "slower than shit." Business cards and brochures proved no more useful than telegrams.

And then everyone had an excuse. Some were too busy to meet, which just seemed cruel. (SAM could help with that!) One firm said bricks laid by SAM were *too perfect* and would necessitate tearing out all of the human-laid bricks on a job. Mostly, though, it came down to habit. People weren't exactly complacent; they were just . . . stuck.

Scott, realizing it took many attempts to get unstuck, figured his best chances were with firms that had already tried using SAM, even if those trials had gone poorly, even if the firm's men hated using the machine. In addition to reapproaching Soderberg (of the Laramie job), Scott had also reapproached Fernandes Masonry (of the Lunenburg job, which Zak had once vowed to strike from CR's company history), and even Cascade Masonry (of the devastating Seattle job). Indeed, Scott had wanted to salvage his company's reputation before rumors spread too far, but he also wanted work, even when he was pretty sure it wouldn't lead to a sale. He was desperate. He'd even considered modifying SAM so that the machine could rebuild the brick ovens of a coke processor.

When new companies nibbled, it was impossible to distinguish between intrigued parties and interested ones. Everyone, it seemed, wanted to ogle SAM. But that was usually all. Before anybody at CR could figure out a new party's stance, their guys came to Victor for a demo. CR performed so many demos out back by the highway (or inside, if weather demanded) that, by the spring of 2016, Scott's engineers were complaining about them. The demos, they said, were taking them away from the actual engineering they were trying to do. Couldn't people just watch videos of SAM online? They seemed to be suffering from a marketing vacuum, which was strange, because Eric Ries had specifically discounted the concern.

When a demo led to a rental, that still left Construction Robotics in a precarious financial state. Until CR could hand off the machine and be confident it wouldn't get bitten in the ass, the company provided an operator in the mold of Stephen or Ken. It was like renting a car in India: A driver was part of the package. While the robot consumed propane and slept on the scaffold, the human attendant consumed food and slept in a hotel and traveled between these locales with a rental car, which ate away at CR's profit margin. Throw in the cost of flights home, and nothing remained. It was doubly demoralizing because, by way of further excuses, these rentals hadn't led to sales, and CR needed six sales a year just to break even. At the helm, Scott wasn't even treading water.

Usually, masonry firms said they didn't have enough jobs to keep SAM busy. (Scott always countered with the argument that with SAM, a masonry firm could competitively bid for a lot more masonry work.) They also said SAM's price tag, three times higher than a screed or forklift, was hard to swallow. Regarding that price tag, Zak and Scott and Nate and Chris Raddell had analyzed the masonry market and gathered that SAM made an attractive proposition in Rochester and Syracuse, but also in Chicago, Pittsburgh, Cleveland, Columbus, Cincinnati, Dayton, Toledo, and St. Louis if it laid one thousand bricks a day. Unfortunately, most of these places were not within a four-hour drive of Victor, a zone deemed a reasonable striking range. Down south, where wages were lower, the machine needed to lay more like twelve hundred. How had their assessment been so wrong?

And so, while the company burned $1.5 million a year developing SAM, it pulled in at best 2 or 3 percent of that. Scott, all the while, was making no more money at the helm of CR than he had his first year out of college. Talk about a lean start-up. It didn't help that the masonry contractor who'd rented SAM in Virginia and D.C. was charged by the Department of Justice with failing to pay nearly $1 million in employment taxes, and went poof before Scott was paid for either job. Nor did it help that the tab for the Massachusetts job went uncollected as well.

The pace at which things proceeded left Scott demoralized, because his ambitions extended far beyond SAM and encompassed most of the construction industry: He wanted to develop concrete-grinding machines and brick-cutting machines and spraying machines and trash-picking machines and a few others, all working off the same robot base. He was sure that the map tool alone, just one component of SAM, held so much possibility! At various points, Scott considered licensing it to architects to pull in revenue, or distributing the software for free, toward creating a new industry standard. He was sure he was sitting on so much potential, and he was tired of waiting to break through.

If there was any consolation, or maybe just irony, it was that amid all of the fruitless pitching, which itself took place amid a torrent of difficult jobs, Zak bought and began renovating a foreclosed house around the corner from Scott and discovered that construction management was as complicated and frustrating at home as it was on any given commercial jobsite. To upgrade his kitchen and upstairs bathroom, Zak hired subcontractors, and to his dismay, these subcontractors disappointed him. Their speed and quality and price were not satisfactory, a situation that was all too familiar.

Scott had once considered selling SAM at a loss, on the condition that for every brick SAM placed, CR would collect a royalty of five or ten cents—but this idea withered away, partly because CR didn't have enough cash in hand to see the strategy through (payback would have taken a decade), and partly because he thought firms liked to own their equipment outright. And so while Scott sought out the elusive magic sales formula, he asked the company's advisory board for help. He went so far as to tell the board that his company sucked at sales—which was a dig at Chris Raddell. In Scott's words, Chris Raddell had a flurry of words but not of dollars.

Zak, too, had his doubts about Chris Raddell. The way Zak saw it, he could open doors but not close deals. He seemed unable to remember the company line, didn't know when to talk and when to listen, was past his prime, and—because he spent way more time on the phone or on the road than at actual jobs—always overpromised. Zak blamed him for lining up the Fernandes job and failing to provide adequate warning. "We never should have hired him," Zak said.

But Chris Raddell knew as well as anyone the conservatism that resulted from having lost your shirt in the recession, and he knew that his words amounted to nothing without numbers to back them up. "I can tell you my car's faster than yours all day long," he said by way of a metaphor. "Take me to the track. Take me to the track." If anything, his salesmanship was worthy, while the product he was pushing was not. He'd just been hired too early. Besides, his heart was decent. Alone among the engineers at CR, Chris Raddell had attended Scott's brother's funeral.

As Chris Raddell's sales tactics began to wear thin, Zak developed his own approach. Rather than playing the educated businessman spewing data and logic and reason, he adopted the role of a down-to-earth, unthreatening guy. At World of Concrete the second year, he handed out cans of beer, hoping to earn goodwill among the parade of masons streaming by to have a look-see. He'd also hoped to give out T-shirts—he knew guys loved swag, and the marketing would have been priceless. Zak's mistake, though, had been throwing some last-minute ideas in front of CR's T-shirt design committee, which was comprised entirely of engineers. One design idea for the front of a shirt, the "evolution of a bricklayer," showed the path from caveman to bricklayer to robot—but Zak quickly realized that, for all its creativity, the message was exactly what CR did not want to promote. Another idea, the "robot cowboy," featured a figure twirling a lasso atop the robot. It was never clear if this made the robot a bucking bronco or, à la Strangelove, a nuclear weapon, but it got shot down regardless. Someone suggested a woman riding the robot, knowing that would draw eyes, but it was rejected promptly. Zak abandoned the figure and suggested an iteration involving CR's logo and only six words: DANGER: DO NOT RIDE THE ROBOT!

Kerry—CR's oldest employee—thought it was over-the-top. The head of engineering was confused by the word "danger." Stephen was also averse to the word. He preferred something gentler, as in: "No, you may not ride the robot." Glenn agreed; he said a neutral statement wouldn't irritate people already threatened by SAM. John, in rare alignment with Zak, thought the design was fine. Switch "danger" to "warning," he said, and get on with printing a thousand of them already. Chris Raddell, in charge of sales, said not so fast. He wanted the shirts to feature a graphic of the Stäubli arm, and highlight some "value-added attribute points," namely safety, productivity, and quality. Zak thought that would have been the stupidest shirt he'd ever seen. But Tim agreed with Chris Raddell. He said conservative was the way to go. And so, because Zak tried to be inclusive, and because he deferred to Chris Raddell, no T-shirts were ordered. His marketing approach withered away.

The result was that Construction Robotics had no easy way to deflect the barrage of insults thrown at SAM in Las Vegas. Spectators, who stood with crossed arms and shook their heads as SAM built short demo walls, heckled like Bostonians. They called SAM bullshit and asked if the machine had a higher gear. "He ain't gonna win the Bricklayer 500," one man said. Another said the machine needed to be sabotaged. Still another asked where the gas tank was, so he could pour some sugar in it. The harshest were the father and son from Australia.

"It won't work," the father kept saying.

"It just did, right there!" the son kept saying.

But there was no convincing the father. "Don't worry," he said condescendingly to Glenn, who was manning the controls of the two-ton bricklaying robot. "Someday you'll learn to be a bricklayer."

The only respite was when the emcee for the Bricklayer 500, live on the Jumbotron, in front of four thousand people, called SAM amazing and impressive, and said SAM's walls were "absolutely gorgeous." To Zak, the emcee was a better salesman than Chris Raddell.

Scott told CR's advisory board that he was sure SAM could kick ass, as the machine occasionally had. He was so certain that he fantasized about starting his own masonry company, which would allow him to bypass the unholy mess that was construction, and—unburdened by interference and inexperience and incompetence—show the world that his machine could crank. In that way, CR could grow into the biggest masonry contractor in the country. If his company followed such a course, he could embark on a complete redesign, make SAM a third the size and weight, and incorporate Leica's laser tracker, allowing him to ditch the laser box, the poles, the mounts, the tabs, the red dot, the beeping, the measuring, the Wi-Fi—all of it. In this way, he was sure he could get SAM's cycle time down to six seconds, run at world-record pace all day, and never take shit from anybody. He'd prove everybody wrong.

This, the board had agreed, was untenable. CR was a technology company, staffed with pale, skinny engineers who sat in front of computers. To become a masonry firm was more than a mere pivot. The board understood that SAM was facing a headwind of ingrained habit and complacency. Change across one of the country's biggest industries did not come about overnight. But Scott was already a decade into things, and he knew that most start-ups, like most restaurants, failed in only a few years. To all present, the matter was one of perseverance—a term Eric Ries disparaged. The phrase "live to fight another day" came up. Toward this end, they discussed hitting up CR's original investors a second time (even though they lacked a compelling success story), selling equity in the company, seeking funding from Husqvarna or Volvo or billionaires in Rochester or even dreaded venture capital firms. In the end, CR pursued a half-million-dollar loan so that CR could build six more machines and start renting them, too.

But Scott had never aspired to rent machines. The rentals he negotiated from November 2014 through November 2016 were merely the method by which he could get the stories he needed to show the machine's worth. A worthy machine could sell like hotcakes. But without the stories, he was no better than so many of the overslick entrepreneurs in Silicon Valley, blurting out hype and guesses in equal proportion. Proven examples would validate his vision. Bricks would speak for themselves. Once he had stories, he wanted to sell two hundred machines a year.

The board had other ideas. They saw more wisdom in renting machines until the tide turned. Renting established a market and ensured steady income in economies good and bad. Hydro-Mobile had invaded the frame-and-brace scaffold market by renting. To Scott (and to Don Golini), the rental approach intuitively registered as pulling defeat from the jaws of victory. Renting machines was prudent; selling machines was commanding—and Scott wanted to show command.

Nearly a year earlier, at the end of 2015 and on the heels of the Brawdy job, CR had actually sold two machines (at a discount) to an equipment distributor in St. Louis named Irwin Products. For a moment, this had felt like a victory to Scott—and then the feeling quickly faded, leaving an eerie sense of liability. Astutely, Scott foresaw that this distributor, for all the motivation he provided (per MVP dictate, his requests were sure to drive progress), would also present a headache. Worse, the "sale" didn't prove SAM's worth, because Vince Irwin would just be renting his bricklaying machines to all the various masonry contractors he'd come to know in his many years in the industry. He'd be renting machines temporarily to firms unwilling to commit—which was no more than CR had been able to pull off. The only difference was that Vince Irwin, whose grandfather had built the St. Louis Armory, and whose father had built the Edison Building, and who himself had built St. Luke's, could leverage his established reputation in the trade.

For a long time, Scott and Nate (and Zak and Chris Raddell) thought Clark would be the first buyer. Even though Clark didn't perform its own masonry, there was a chance, according to its CEO, that Clark would establish a masonry branch based around SAM. In time, CR's numero uno prospect became Brawdy. For months, Scott thought Brawdy was on the hook—and on his birthday, he all but groveled as he tried to sell Jim a machine for less than half its retail price. Then Findorff entered the picture, and Scott and Nate fantasized about

Findorff until Zak had two meetings in Boston. One of them was with the president of Suffolk, the third-largest GC in the country, and one was with the brothers at the helm of Consigli, a $1.5 billion construction firm. Both said they'd use SAM; the latter said the likelihood was 100 percent. Boston cast a shadow all the way to Madison, until it faded away. Then Scott met David Hux, whose passion for the robot was the first to rival his own. He was like an eighth-grade boy who bumped into a girl in the hall and immediately conjured their eventual marriage.

Before anything was consummated, Wilhelm showed up. Wilhelm—which already had Hydro-Mobiles, and already provided incentives for masons to reach big numbers, and already had in place a honed system for managing the choreographed chaos of construction—actually committed. The purchase order came along at the crucial moment. It wasn't enough to flip CR's finances around, but it suggested that commercial success was possible.

By the next spring, SAM could lay a brick every 8.2 seconds, and do so regularly, thanks to a refined buttering path. Where once the machine had altered its buttering depending on which way it was rolling, now the engineers had it butter the same way—head to bed—whether it was rolling left or right. They did this because the grab-lower-pass-twist-extend-drop-unwind was that much faster, and who cared if a dry brick hit a glob of mud instead of a glob of mud hitting a dry brick? The motion was damn near perfect.

By then, the machine could also lay soldier (upright) courses. The laser box was new. The way the laser box moved up and down the pole was improved. The new fold-down track was so slick it made their original welded-on track look like bamboo. For four consecutive days on a job in Virginia, SAM laid upward of 2,500 bricks, including one day when the machine put down 3,261 bricks in a single shift. SAM broke as many records as Katie Ledecky.

In the second half of 2017, SAM really started cranking. One machine laid a hair under 10,000 bricks in three days. Two days in a row, masons averaged 385 bricks per hour all day. Sometimes they hit 390.

Showing off, masons worked a twelve-hour day and got SAM to lay 3,851 bricks. Once, in less than ten hours, SAM placed 3,000. In Roanoke, Virginia, after seventy-five days of running, SAM finished the first of the tall, broad, windowless walls—"money walls" if ever there were any—on the fifteen-story Poff Federal Building. In that time, the machine had put down 129,337 bricks.

There was another indication that SAM had arrived where Scott always wanted: On just one job, the machine laid more bricks than it had on most of the dozen previous jobs *combined*. It had taken three long years—one when they learned how to run the machine, one when they taught others how to run it, and another when they refined it—but the endeavor hadn't been the impossible task that Bruce Schena had declared it only a generation before. Scott had not only proceeded as boldly as Coach (without the persona) and as persistently as a distance swimmer but as stubbornly as a bricklayer.

By then, Nate's first inkling of an improved way to lay the building blocks of civilization was two decades old. He'd turned seventy. He soon retired from Hueber-Breuer but kept an office there, from which he continued to work full-time on SAM.

At the start of 2018, the scrappy little start-up moved beyond the orbit of PMD and into a new, larger office. Kerry Lipp, Scott's second hire, had quit, as had two other engineers (including the director of engineering) and the office manager. Scott hired more engineers, as well as another operator and a sales director, bringing the number of mouths to nearly two dozen. Somehow Zak ended up above the shitter again, but at least he wasn't in charge of marketing anymore. Not that it made much of a difference: One-upping the *Rochester Democrat and Chronicle*, FOX News announced that SAM could lay a thousand bricks per hour. As if to compensate, the *New Yorker* said SAM laid only a thousand a day. And then *USA Today*, in two very legible photo captions, called the bricklaying robot SAMANTHA. That was straight-up defamation.

Of the original crew, Mike Oklevitch, Glenn, Tim, Kim, and Chris Johnson remained—although these last three would decamp before the year was over. Up until they left, they often labored as they had in the

Rocky era. Upgrading SAM, they worked weekends, sometimes stayed until midnight. The engineers designed new poles free of the annoying and expensive tabs, and new track that fit Fraco and Bennu scaffolds. In this way, they whittled away at the excuses that masonry firms came up with for not trying and buying SAM. Scott, meanwhile, landed two one-year leases, which showed a certain faith in the capability of the bricklaying machine that looked like nothing so much as a hot dog cart.

But still, Construction Robotics did not sell a machine to another mason contractor.

On the other hand, at one time CR had five SAMs running at once, such that the company, all told, laid four thousand bricks before lunch. Slowly, Construction Robotics modernized the construction industry. Scott, of course, wanted more: He wanted a few SAMs, each working in tandem with a Leica laser tracker, in every big city in America, and CR's value on a trajectory toward a billion dollars.

The engineers also built the block-laying machine that they'd nearly devised at a crucial junction five years before, and which masons on two continents had been practically begging for. They called it MULE, for "material unit lift enhancer," and it was better than MAMA, even if it was less a robot than a counterbalanced gripper. With it, men could place forty-pound blocks at almost the rate generally associated with bricks. CR listed the machine for seventy-three thousand dollars, and immediately, orders came pouring in. By Memorial Day, CR had sold seventy-five.

All of a sudden, CR's revenue situation improved dramatically—which suited Scott, because already, his dry-erase board was covered with years' worth of inventions yet to be developed.

Business was as exciting as ever, and only a smidge less demanding, but Scott found a way to abandon all of the frustration and relax. Having bought an old run-down cottage on the south shore of Lake Ontario for about the price of SAM's Stäubli, he fixed it up. It wasn't much more than a brown double-wide trailer, but it was under fifty yards from the water and only a few doors down from his parents' cottage, where he'd once sailed with his brother. Among other things, he installed a cell-phone booster and a desk, so that he could work before a window while absorbing a full panorama of the line where water met sky. On the horizon to the northeast, he watched boats go by. Sometimes he brought his kids and set them free to run around the neighborhood as kids of so many previous generations—before helicopter parents and extracurriculars and iPads—had done. Such days registered deeply in him as amazing and dreamlike. The water exerted a calm over the engineer.

Then the engineer came up with an idea. He wanted to build a dock—a special dock. He'd never built a dock, but that had never stopped him. With some help from a welder, he built a sixty-foot-long dock like no other. Thanks to electric motors, he could adjust the height of the whole thing with a button, so that when the water was flat and glassy, he could walk level with the surface, and when the water was choppy, he could walk right over the waves.

Acknowledgments

For first encouraging me to check out World of Concrete, where the idea for this book was born, I tip a grubby hat to Brian Carney. For supporting works that promote the public understanding of science, I brush off the crud and bow gratefully to Doron Weber and the Alfred P. Sloan Foundation.

For being there, in the construction zone, thanks to Abe Streep, Ben Berk, Ann and Evan Smith, Brian Feldman, Rob Gorski, and the Botkin clan.

For examining the blueprints, thanks to Rinker Buck, Evan P. Schneider, Justin Neuman, James Somers, John Shea, David Baron, Erin Espelie, Dan Becker, Nick Masson, Michael and Jess Cody, and Erin Fletcher.

For shoring up teetering walls, I'm grateful to Andrew Green, Clif Wiens, Adam Amir, Leah Goodman, Tasha Eichenseher, and of course, Mom and Dad.

For superintending this wacky project and seeing it built right, thanks to Richard Morris at Janklow & Nesbit, and Jofie Ferrari-Adler and the gang at Simon & Schuster.

Most of all, thanks to Scott, Zak, and the rest of the Peterses and Podkaminers for letting me in, trusting me, and treating me like family.

About the Author

JONATHAN WALDMAN, who has held journalism fellowships from the Scripps Foundation and the Alicia Patterson Foundation, has written for publications ranging from the *New York Times* to *McSweeney's*. His first book, *Rust: The Longest War*, was a finalist for the Los Angeles Times Book Prize. A native of Washington, D.C., he graduated from Dartmouth and lives in Colorado.